Methods for Mining and Summarizing Text Conversations

Synthesis Lectures on Data Management

Editor
M. Tamer Özsu, ***University of Waterloo***

Synthesis Lectures on Data Management is edited by Tamer Özsu of the University of Waterloo. The series will publish 50- to 125 page publications on topics pertaining to data management. The scope will largely follow the purview of premier information and computer science conferences, such as ACM SIGMOD, VLDB, ICDE, PODS, ICDT, and ACM KDD. Potential topics include, but not are limited to: query languages, database system architectures, transaction management, data warehousing, XML and databases, data stream systems, wide scale data distribution, multimedia data management, data mining, and related subjects.

Methods for Mining and Summarizing Text Conversations
Giuseppe Carenini, Gabriel Murray, and Raymond Ng
2011

Probabilistic Databases
Dan Suciu, Dan Olteanu, Christopher Ré, and Christoph Koch
2011

Peer-to-Peer Data Management
Karl Aberer
2011

Probabilistic Ranking Techniques in Relational Databases
Ihab F. Ilyas and Mohamed A. Soliman
2011

Uncertain Schema Matching
Avigdor Gal
2011

Fundamentals of Object Databases: Object-Oriented and Object-Relational Design
Suzanne W. Dietrich and Susan D. Urban
2010

Advanced Metasearch Engine Technology
Weiyi Meng and Clement T. Yu
2010

Web Page Recommendation Models: Theory and Algorithms
Sule Gündüz-Ögüdücü
2010

Multidimensional Databases and Data Warehousing
Christian S. Jensen, Torben Bach Pedersen, and Christian Thomsen
2010

Database Replication
Bettina Kemme, Ricardo Jimenez Peris, and Marta Patino-Martinez
2010

Relational and XML Data Exchange
Marcelo Arenas, Pablo Barcelo, Leonid Libkin, and Filip Murlak
2010

User-Centered Data Management
Tiziana Catarci, Alan Dix, Stephen Kimani, and Giuseppe Santucci
2010

Data Stream Management
Lukasz Golab and M. Tamer Özsu
2010

Access Control in Data Management Systems
Elena Ferrari
2010

An Introduction to Duplicate Detection
Felix Naumann and Melanie Herschel
2010

Privacy-Preserving Data Publishing: An Overview
Raymond Chi-Wing Wong and Ada Wai-Chee Fu
2010

Keyword Search in Databases
Jeffrey Xu Yu, Lu Qin, and Lijun Chang
2009

Methods for Mining and Summarizing Text Conversations

Giuseppe Carenini, Gabriel Murray, and Raymond Ng

www.morganclaypool.com

ISBN: 9781608453900 paperback
ISBN: 9781608453917 ebook

DOI 10.2200/S00363ED1V01Y201105DTM017

A Publication in the Morgan & Claypool Publishers series
SYNTHESIS LECTURES ON DATA MANAGEMENT

Lecture #17
Series Editor: M. Tamer Özsu, *University of Waterloo*
Series ISSN
Synthesis Lectures on Data Management
Print 2153-5418 Electronic 2153-5426

Methods for Mining and Summarizing Text Conversations

Giuseppe Carenini, Gabriel Murray, and Raymond Ng
University of British Columbia

SYNTHESIS LECTURES ON DATA MANAGEMENT #17

MORGAN & CLAYPOOL PUBLISHERS

ABSTRACT

Due to the Internet Revolution, human conversational data—in written forms—are accumulating at a phenomenal rate. At the same time, improvements in speech technology enable many spoken conversations to be transcribed. Individuals and organizations engage in email exchanges, face-to-face meetings, blogging, texting and other social media activities. The advances in natural language processing provide ample opportunities for these "informal documents" to be analyzed and mined, thus creating numerous new and valuable applications.

This book presents a set of computational methods to extract information from conversational data, and to provide natural language summaries of the data. The book begins with an overview of basic concepts, such as the differences between extractive and abstractive summaries, and metrics for evaluating the effectiveness of summarization and various extraction tasks. It also describes some of the benchmark corpora used in the literature.

The book introduces extraction and mining methods for performing subjectivity and sentiment detection, topic segmentation and modeling, and the extraction of conversational structure. It also describes frameworks for conducting dialogue act recognition, decision and action item detection, and extraction of thread structure. There is a specific focus on performing all these tasks on conversational data, such as meeting transcripts (which exemplify synchronous conversations) and emails (which exemplify asynchronous conversations). Very recent approaches to deal with blogs, discussion forums and microblogs (e.g., Twitter) are also discussed.

The second half of this book focuses on natural language summarization of conversational data. It gives an overview of several extractive and abstractive summarizers developed for emails, meetings, blogs and forums. It also describes attempts for building multi-modal summarizers. Last but not least, the book concludes with thoughts on topics for further development.

KEYWORDS

automatic summarization, abstraction, extraction, conversations, text mining, sentiment, subjectivity, topic modeling, evaluation, emails, weblogs, meetings, chats

To our families,
Cristina, Elisa, Francesca, Hamish, Heather, Michele, and Kevin

Contents

1 Introduction **1**

1.1 The Rise of Text Conversations 2

1.1.1 The Internet Revolution 2

1.1.2 The Speech Technology Revolution 5

1.2 Application Scenarios 6

1.3 Related Topics and Background Readings 9

1.4 Mining and Summarizing Text Conversations: An Overview 9

1.4.1 Mining Text Conversations 11

1.4.2 Summarizing Text Conversations 13

1.5 Book Preview 19

2 Background: Corpora and Evaluation Methods **21**

2.1 Corpora and Annotations 21

2.1.1 Meeting Corpora 22

2.1.2 Email Corpora 25

2.1.3 Blog Corpora 27

2.2 Evaluation Metrics for Text Mining 27

2.2.1 Precision, Recall and F-Scores 28

2.2.2 ROC Curves 28

2.3 Evaluation Metrics for Summarization 32

2.3.1 Intrinsic Summarization Evaluation 32

2.3.2 Extrinsic Summarization Evaluation 37

2.3.3 Linguistic Quality Evaluation 38

2.3.4 Evaluation Metrics for Summarization: A Final Overview 39

2.4 Conclusion 40

2.5 Important Points 40

2.6 Further Reading 41

3 Mining Text Conversations **43**

3.1 Introduction 43

3.2 Topic Modeling: Topic Segmentation and Topic Labeling 43

3.2.1 Topic Modeling of Generic Text 44
3.2.2 Topic Modeling of Conversations 49
3.3 Sentiment and Subjectivity Detection 55
3.3.1 Sentiment Detection Background 55
3.3.2 Sentiment Detection in Conversations 58
3.4 Conversational Structure 60
3.4.1 Unique Features of Human Conversations 60
3.4.2 Dialogue Act Modeling 62
3.4.3 Decision Detection 68
3.4.4 Action Item Detection 69
3.4.5 Extracting the Conversational Structure 72
3.5 Conclusion 76
3.6 Important Points 76
3.7 Further Reading 77

4 Summarizing Text Conversations 79
4.1 Introduction 79
4.2 Summarization Framework and Background 81
4.2.1 Assumptions and Inputs 82
4.2.2 Measuring Informativeness 84
4.2.3 Outputs and Interfaces 85
4.3 Summarizing Conversations in One Domain 86
4.3.1 Summarizing Emails 86
4.3.2 Summarizing Meetings 91
4.3.3 Summarizing Chats and Blogs 94
4.4 Summarizing Multi-Domain Conversations 97
4.4.1 Abstractive Conversation Summarization: A Detailed Case Study 99
4.5 Conclusion 101
4.6 Important Points 102
4.7 Further Reading 102

5 Conclusions / Final Thoughts 103

Bibliography 107

Authors' Biographies 119

CHAPTER 1

Introduction

Before the invention of the Internet and the creation of the Web, the vast majority of human conversations were in spoken form, with the only notable, but extremely limited, exception being epistolary exchanges. Some important spoken conversations, such as criminal trials and political debates (e.g., *Hansard*, the transcripts of parliamentary debates), have been transcribed for centuries, but the rest of what humans have been saying to each other, throughout their history, to solve problems, make decisions and more generally to interact socially, has been lost.

This situation has dramatically changed in the last two decades. At an accelerating pace, people are having conversations by *writing* in a growing number of social media, including emails, blogs, chats and texting on mobile phones. At the same time, the recent, rapid progress in speech recognition technology is enabling the development of computer systems that can automatically transcribe any spoken conversation.

The net result of this ongoing revolution is that an ever-increasing portion of human conversations can be stored as *text* in computer memory and processed by applying Natural Language Processing (NLP) techniques (originally developed for written monologues - e.g., newspapers, books). This ability opens up a large space of extremely useful applications, in which critical information can be mined from conversations, and summaries of those conversations can be effectively generated. This is true for both organizations and individuals. For instance, managers can find the information exchanged in conversations within a company to be extremely valuable for decision auditing. If a decision turns out to be ill-advised, mining and summarizing the relevant conversations may help in determining responsibility and accountability. Similarly, conversations that led to favorable decisions could be mined and summarized to identify effective communication patterns and sources within the company. On a more personal level, an informative summary of a conversation could play at least two critical roles. On the one hand, the summary could greatly support a new participant to get up to speed and join an already existing, possibly long, conversation (e.g., blog comments). On the other hand, a summary could help someone to quickly prepare for a follow-up discussion of a conversation she was already part of, but which occurred too long ago for her to remember the details. Furthermore, the ability to summarize conversations will also be crucial in our increasingly mobile world, as a long incoming message or an extensive ongoing conversations could be much more easily inspected on a small screen in a concise, summarized form.

This book presents a set of powerful computational methods to mine and summarize text conversations, where a text conversation is either one that was generated in writing, or one that was originally spoken and then automatically transcribed. Different kinds of useful information can be mined. We will describe how to detect what topics are covered in a given text conversation, along with what opinions the conversation participants have expressed on such topics. We will also discuss how the underlying structure of a text conversation can be determined by identifying specific dialogue acts (e.g., request, answer) and their relationship (e.g., a question/answer pair). All these kinds of extracted information, expressed in term of topics, opinions and conversational structure, can then be used to summarize the conversation. We will see how different summaries can be generated at different levels of granularity depending, for instance, on the audience and specific information needs.

Most of the studies we cover in this book involve techniques to deal with text conversations in a particular modality (e.g., extracting opinions from an email conversation). However, we will also discuss more recent work which is increasingly tackling the challenges of mining and summarizing conversations spanning multiple modalities (e.g., a transcript of a meeting that was followed up by an email conversation).

It is widely accepted that we are in the midst of an epochal revolution in how people communicate. We believe that researchers have just started to envision the plethora of opportunities presented by the widespread availability of text conversations. We hope that our readers, after learning the various techniques described in this book for mining and summarizing text conversations, will be able to find new and creative ways to apply them, making (computer-mediated) human communication ever more effective.

In the remainder of this introduction, we first present some data on the amazing pace at which human conversations are moving from the spoken to the text form. Then, we discuss some key application scenarios for conversation mining and summarization. After that, we will conclude with an overview of the research space for the computational techniques we will explore in this book.

1.1 THE RISE OF TEXT CONVERSATIONS

In this section we first consider the rise of internet technologies and subsequently discuss contemporaneous developments in speech technology, and describe how both have transformed the way we communicate with one another.

1.1.1 THE INTERNET REVOLUTION

The rapid adoption rate of new, Web-based forms of communication, has beaten even the most hyperbolic predictions. Email was born in 1971 and is considered the grandparent of all Web-based social media. Today, it is used daily by billions of people all over the world in a seemingly unlimited variety of communicative settings. We email our friends to organize a trip over the weekend, our colleagues to discuss next year's budget, and a car dealer to bargain the price of a new car. As Baron [2008], a leading expert in Computer Mediated Communication, put it: "Emails have style and content as diverse as people using it". Also, email is clearly a domesticated technology, i.e., it has become a normal component of daily living [Baron, 2008]. Although precise data are always difficult to come by, a quick Web search tells us that in 2009 Yahoo Mail had a billion emails passing through its servers every day (as reported by its CEO, Carol Bartz). According to more recent data on websites weekly visits[1], MS Live Mail and Gmail, the other two top email providers, generate comparable traffic (a third of Yahoo each). The time people spend on emails has also been growing. Already in 2004, in a survey by Ferris Research with 840 U.S. businesses[2], 10% of their workforce spent more than half the workday (4-plus hours) on email, and 86% engage in personal email correspondence. Nowadays, even if email has to compete with other social media, according to a 2010 survey by Nielsen[3], it is still the third most frequent activity of U.S. Internet users (see Figure 1.1), and the top one in mobile Internet activities, at 40%.

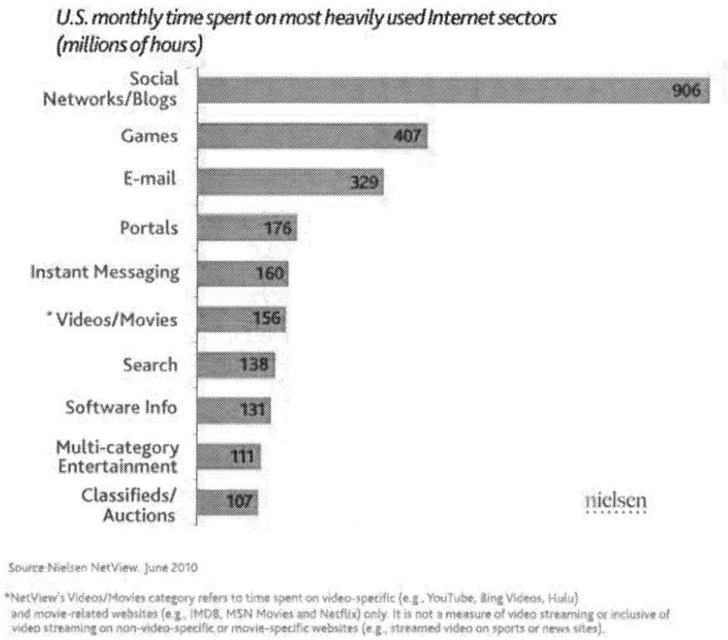

Figure 1.1: Internet usage in the U.S., 2010.

Another Web-based form of communication closely related to email is instant messaging (often referred to as chats), whose origins also date back a few decades. The key difference between emails and instant messaging is that email is asynchronous, while instant messaging is synchronous. If you send an instant message, you expect a reply right away, as you would in a face-to-face conversations. In contrast, it may be acceptable if an email is answered days later. Instant messaging has also witnessed a staggering growth. According to Wikipedia, the number of people with instant messaging accounts was estimated in 2006 to be 340 million worldwide. Currently, this number is speculated to be approximately 1 billion.

synchronous

asynchronous

[1] http://www.hitwise.com/us/datacenter/main/dashboard-10133.html
[2] http://www.ferris.com/category/topics/statisticssurveys/
[3] http://goo.gl/zsnw5

The term Blogs (= Web Logs) was coined in 1997 and since then blogging has quickly become another killer application for Web-based communication. While the domestication of email took many years, blogs became domesticated in a much shorter time [Baron, 2008], probably because usage of the Web was already quite common. Blogging is now a common way for people to freely publish their thoughts about almost any conceivable topic on the Web and to engage in online discussions. These discussions are forms of text conversations. After an initial contribution is posted, for instance a news article, a proposal, question, or review, anyone can comment about it, and all these additional comments can in turn generate possibly long threads of further discussion. Similarly, to emails, the growth of the blogosphere (the space of all blogs) has been astonishing. According to a report published by Technorati in 2008[4] the blogosphere had consistently doubled every 5 months for the preceding 4 years and the size was estimated to be, at that time, approximately 133 million blogs. More recent data come from a new report released by eMarketer in 2010[5]. In that year, 51% of U.S. internet users, or 113 million people, read blogs on a monthly basis. By 2014, the blog audience is expected to rise to 60% of internet users, or 150 million people. The number of bloggers was also predicted to grow, though somewhat more modestly. In 2010, 11.9% of U.S. internet users keep blogs. By 2014, there will be 33.4 million bloggers in the U.S., representing 13.3% of internet users. Notice that the numbers from this report are underestimates, as eMarketer counts only people who blog, excluding marketers or media companies with public-facing blogs.

In spite of the phenomenal growth of email, instant messaging and blogs, it seems that new forms of social media, where people can engage in text conversations, are being constantly created. And their rate of domestication becomes shorter and shorter. For some users, Twitter and Facebook

micro-blogs

tweets

have become the social media of preference. Twitter, a site for micro-blogging, also called the "SMS of the Internet"[6], was launched in 2006. It currently has an estimated 200 million users[7], generating 65 million tweets a day[8] and handling over 800,000 search queries per day. Tweets are short messages of up to 140 characters that are often used in online text conversations[9].

Facebook, launched in 2004, is the undisputed leading social network, with 150 million users in the U.S. and half a billion worldwide. In Facebook, people constantly engage in conversations by sending messages to their friend (an email-like service), talking via Facebook chat (an instant messaging service), and writing on their personal walls.

A very recent survey by the Pew Research Centre[10] asked people to indicate which activities they took part in online. Figure 1.2 shows those activities that are partly or entirely of a conversational, social nature. While email is clearly an established part of most people's lives, it is also true that other

[4] http://technorati.com/blogging/state-of-the-blogosphere/

[5] http://www.emarketer.com/Report.aspx?code=emarketer_2000708

[6] SMS stands for *short messaging service*

[7] http://www.bbc.co.uk/news/business-12889048

[8] http://blog.twitter.com/2010/06/big-goals-big-game-big-records.html

[9] 38% of tweets are conversational according to a 2009 study by the market research firm Pear Analytics www.pearanalytics.com/.../Twitter-Study-August-2009.pdf]

[10] http://www.pewinternet.org/Static-Pages/Trend-Data/Online-Activites-Total.aspx

conversation types such as blogs, instant messaging and discussion forums are becoming widely popular.

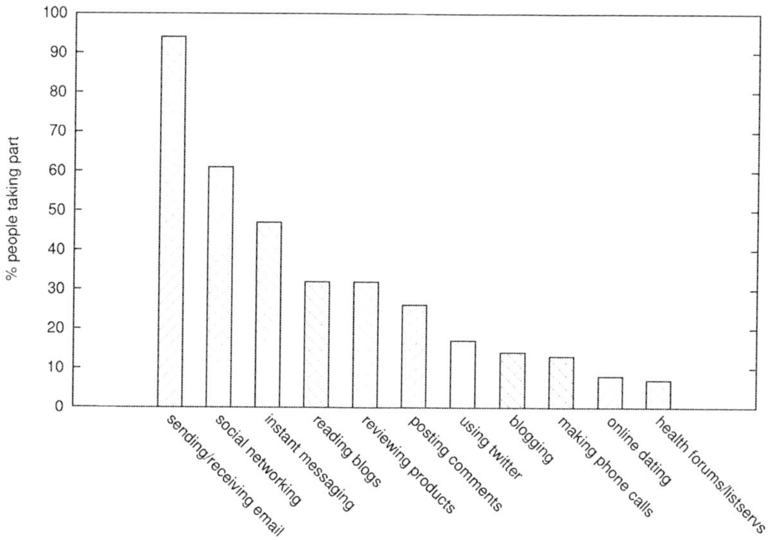

Figure 1.2: Popularity of various online conversational activities.

1.1.2 THE SPEECH TECHNOLOGY REVOLUTION

The rise of text conversations is not only due to the phenomenal adoption of novel Web-based social media; an extremely rapid progress in speech technology is also playing a role. The time is ripe for systems that can automatically transcribe meetings, phone conversations and other spoken interactions. In this book, when we discuss summarization of spoken conversations, we assume the presence of a transcript, which can be either manually written or the output of an automatic speech recognition (ASR) system. That is, we do not consider speech summarizers that work directly and solely off of the speech signal [Penn and Zhu, Forthcoming]. While some speech corpora contain manual transcripts of meetings, lectures or phone conversations, it is more realistic to expect that a deployed speech summarizer would be running on ASR output. And it is owing to the huge advances in ASR that speech summarization has become a feasible and popular area of research in recent years. **ASR**

A survey of ASR advances is far beyond the scope of this book; Jurafsky and Martin [2008] give a clear, accessible overview of modern techniques and historical trends. The biggest advancement began in the 1970s and was popularized in the 1980s with work on statistical speech recognition approaches using Hidden Markov Models (HMMs). Statistical HMM-based systems remain dominant today, but they have been greatly enhanced by increased availability of training data and improved language modeling techniques, among other advancements.

One of the most recognizable speech recognition systems is *Dragon Naturally Speaking* from Nuance Communications, Inc. This software allows users to control their personal computer using voice commands rather than, or in addition to, typed commands. With Nuance's 2010 revenue surpassing $1 billion, it is clear that there is a burgeoning demand for such voice interfaces[11].

WER

A common way of evaluating ASR systems is by measuring the percentage of incorrect words, or the *word error rate* (WER). The WER can vary hugely depending on the task and the environment. One of the simplest recognition tasks is to identify digits spoken in isolation. On this task, state-of-the-art systems feature WERs approaching zero. In contrast, it is much more difficult to recognize continuous speech coming from multiple participants such as in a meeting environment. We will introduce meeting datasets where state-of-the-art recognition systems yield slightly greater than 30% WER. A recurring question, then, is what impact ASR errors have on summarization and text mining tasks.

While Figure 1.2 showed that online conversations are becoming more popular, it is also the case that professionals still spend a great deal of time in meetings. In 2009, Doodle - a company focused on event scheduling - conducted a survey of 2500 administrative and management staff from across Europe and U.S.[12] and found that on average people are attending 7.1 meetings per week, that the meetings last a whopping 2.75 hours each, and with 7 participants in attendance. However, the rise of the Web is also changing the way we meet. Figure 1.3 shows data from the same survey indicating that only around a quarter of these meetings are face-to-face, with many others being conducted online or via conference call. In any event, the average professional spends a great deal of her working life speaking with other people. ASR systems allow us to capture those conversations and feed them into text mining and summarization systems.

1.2 APPLICATION SCENARIOS

The explosive adoption of the new Internet-based social media indicates that they are extremely effective in supporting communication and collaboration. However, we argue that, in several situations, the effectiveness of these new media could be increased considerably by providing users with tools to mine and summarize both past and ongoing conversations. In this section we describe some possible application scenarios. By no means do we claim our list to be complete, and it is one of the goals of this book to foster the creation of novel applications.

- **Join an ongoing conversation:** The government in your country just approved a major policy change. You find an interesting blog/discussion forum about a news article supporting this change with already 50 comments. You strongly oppose the new policy and you would like to present your argument. Should you start a new thread? Or should you contribute to one of

[11] http://www.nuance.com/company/news-room/press-releases/NC_007738
[12] http://www.doodle.com/about/mediareleases/survey.html

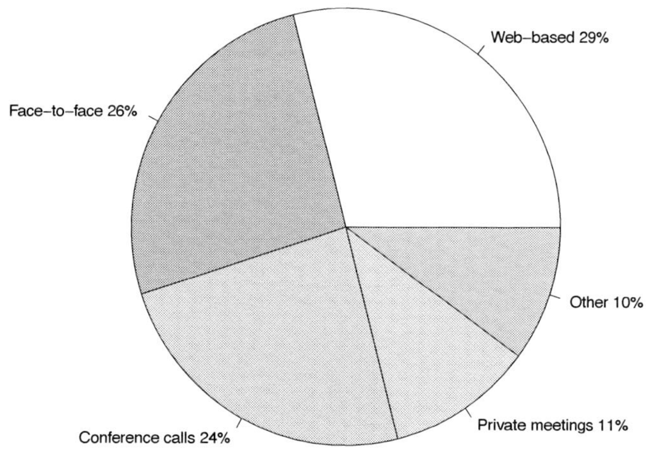

Figure 1.3: The way we hold meetings is being transformed.

the existing threads? To make an informed decision you would like to see a summary of each thread, to know, for instance, whether it is supporting or opposing the change, as well as what are the most frequent arguments used to support/oppose the change.

In a very different setting you manage to join a long meeting only in the middle of it. Before actively joining the conversation you would like to know what was said before. A meeting summarizer could generate meeting minutes on the fly for you.

- **Skim through a long message contributing to an ongoing conversation:** You have just received a long email, contributing to a conversation you have been involved in for several days, along with a few other participants. You may not have time to read the message thoroughly. Yet, you would like to quickly get an idea of the most important new information contained in the message to decide whether you need to reply right away. A summary of the current message, in the context of the previous conversation, would allow you to make this decision more effectively.

- **Business intelligence - Helping preserve corporate memory:** As mentioned previously, mining and summarization technologies can aid an organization carrying out a decision audit in order to improve accountability and hone decision-making practices [Murray et al., 2009]. Such technologies could also be used by human resources staff to gauge employee effectiveness and evaluate communication patterns. Being able to mine and summarize huge amounts of conversational data could also potentially aid the generation of quarterly or annual reports by retrieving data that existed solely in conversations and not in previous reports or formal written documents. A great deal of corporate intelligence exists in informal correspondence that can be translated to more formal documentation.

- **Searching for particular conversations and/or particular messages:** With the phenomenal growth in the amount of text conversations stored in computer memory comes the need for supporting effective search. Generally speaking, the ability to mine and summarize text can benefit any Web search. The more information that can be extracted from text, the more search can be based on the extracted information, rather than on simple matching with the words in the query. For instance, the reader may be very interested in browsing through all the sentences expressing negative opinions, sentences that represent action items, or sentences that describe decisions made. While the underlying conversational data are unstructured, the summary sentences and the linking to original sentence essentially provide structured meta-data to access the underlying data. Moreover, if any document can be effectively summarized, the quality of the presentation of search query results can be improved by presenting a summary as the snippet for each returned documents. Arguably, these advantages would also apply to a search engine for text conversations that relies on the techniques presented in this book. For instance, if it was possible to extract topics and opinions from conversations, a conversational search engine could support queries like: "what messages in the company blogs express opinions on the new budget?". And the output of such search query could be a list of relevant messages summarized in the context of both the query and the conversation (see query focused summarization Section 1.4).

- **Forensic/investigation:** Given the permanent nature of Web-based text conversations, it is not surprising that they have caught the attention of law enforcement organizations as sources of evidence in their investigations. Most countries already accept emails as evidence that can be used in court [Gupta et al., 2004]. For instance, in both the high-profile antitrust trial against Microsoft and the famous Enron scandal investigation, emails were used as evidence in court. In these and similar cases, the amount of data that need to be analyzed is often huge; the Enron email dataset contains about half a million messages belonging to 150 users and stored in 3500 folders. So the ability to mine and summarize the relevant conversations can be highly beneficial.

- **Analyzing large-scale trends:** While many conversations are confined to a small group of friends or colleagues, still others are so large and broad that they effectively feature hundreds of participants making potentially thousands of comments. The growing popularity of Twitter, in particular, has fed this tendency towards large-scale conversation. During a major event such as the Super Bowl or a political uprising in Egypt, relevant Twitter messages (or *tweets*) are sent by the thousands or millions. Some tweets may respond directly to others, while in general the conversation remains vast and amorphous. It is simply not feasible to read all tweets relevant to such a topic, and so mining and summarization technologies can help provide an overview of what people are saying and what positive or negative opinions are being expressed. Sharifi et al. [2010] demonstrate one method of summarizing such large conversations.

1.3 RELATED TOPICS AND BACKGROUND READINGS

While the focus of this book is on summarizing text conversations, there has been considerable work recently on speech summarization. This includes cases where either textual features are supplemented by speech features such as prosody[13] extracted directly from the speech signal, or else textual transcription is bypassed altogether in order to create speech-to-speech summaries. A forthcoming Synthesis Lecture on Speech Summarization [Penn and Zhu, Forthcoming] provides a comprehensive introduction to such methods and represents a nice complement to this one. Another Synthesis Lecture that is very relevant to ours is by Agarwal and Liu [2009]. However, while our focus is mainly on dealing with a single text conversation at a time, they explore approaches for modeling and mining huge collections of intertwined conversations. More specifically, they focus on the space of all blogs that constitutes the blogosphere, and discuss tools for clustering blog conversations, extract communities and identify influential bloggers within a community.

In our exploration of methods and tools for mining and summarizing text conversations, we will often refer to general-purpose techniques for processing and visualizing text. Although we will always try to provide the necessary background, the interested reader can refer to the following publications for a more comprehensive treatment of the different subjects. The leading introduction to the field of Natural Language Processing (NLP) is Jurafsky and Martin [2008], which covers, among others, basic techniques for information extraction, text segmentation and text summarization. Most of the methods presented in this book rely on Machine Learning techniques that have become increasingly popular in NLP in the last decade. All kinds of learning paradigms have been applied to mine and summarize conversations, including supervised, unsupervised and semi-supervised ones. For an introduction to machine learning (ML), see Poole and Mackworth [2010]. For a more comprehensive treatment of the subject, refer to Murphy [expected, Spring 2012]. Semi-supervised methods are described in Zhu and Goldberg [2009]. Opinion mining from text has received a great deal of attention in recent years. Pang and Lee [2008] provide an up-to-date survey of the field. To the best of our knowledge there is no book completely devoted to Information Visualization for text analysis, however a concise introduction to the field is provided by Hearst [2009](Chapter 11).

1.4 MINING AND SUMMARIZING TEXT CONVERSATIONS: AN OVERVIEW

The sample application scenarios we described in Section 1.2 (and many others yet to be explored) require powerful computational methods to mine and summarize text conversations. Although the details of the specific methods will be discussed in later chapters, here we overview the basic intuitions and principles at the core of these methods. Key definitions and illustrative examples will also be provided. As a running example, we will refer to the sample synthetic email conversation shown in Figure 1.4, which involves three participants and seven email messages.

[13]Prosody refers to properties of the acoustic signal associated with an utterance. These include rhythm, stress, and intonation of speech. Prosody has many pragmatic functions. For instance, in many languages, speakers use prosody to convey irony or surprise, to signal emphasis or contrast, and to ask a question.

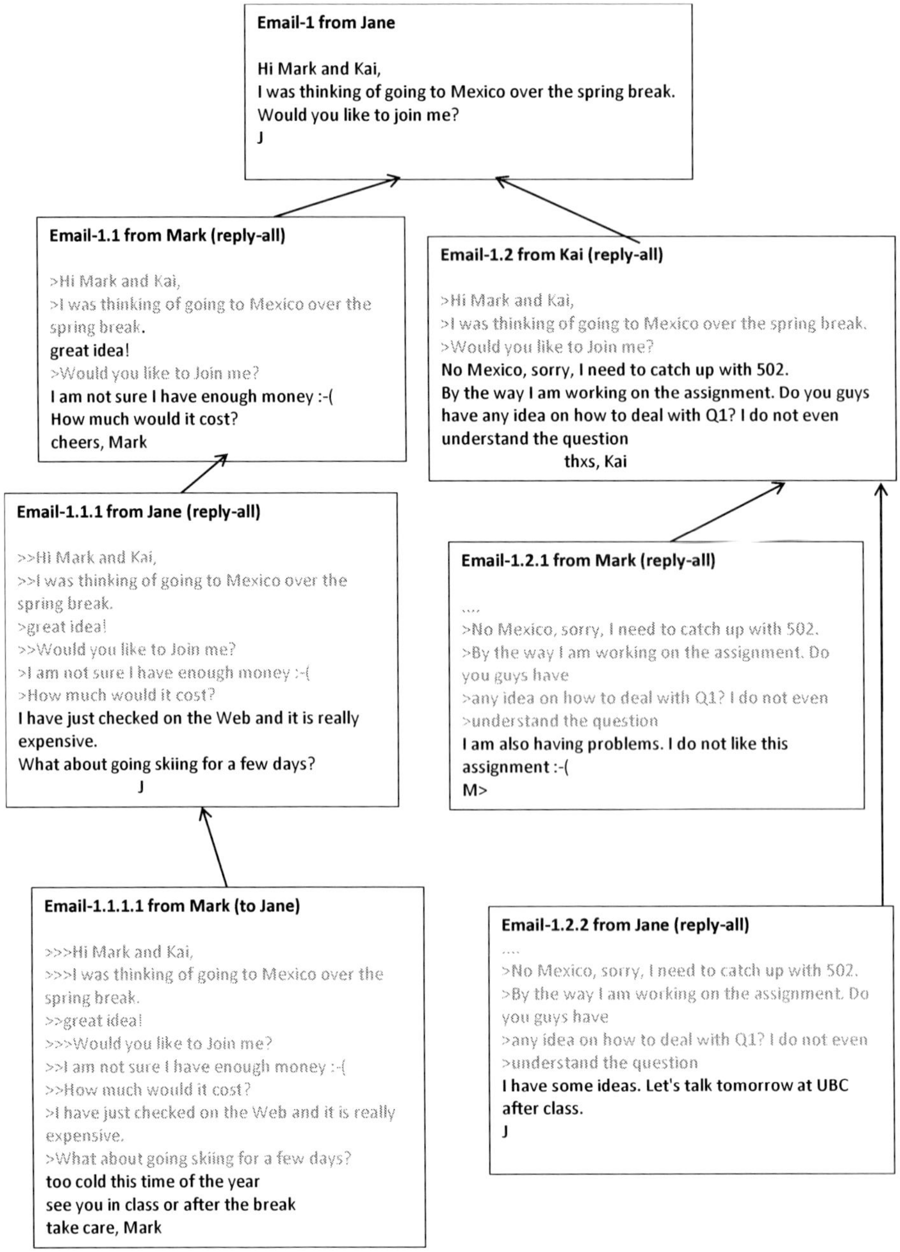

Figure 1.4: Sample synthetic email conversation. The conversation is initiated by Jane (top email) and involves two other participants, Mark and Kai. In each email, new text is in black, while text quoted from preceding emails is in gray. The arrows indicate the reply-to relation.

1.4.1 MINING TEXT CONVERSATIONS

A small set of basic challenging questions can be asked about any text conversations: what topics are covered in the conversation? what opinions do participants express on those topics? what is structure of the conversation? or more specifically, what is the intended function of each particular message (or sentence) and its relationship to other contributions?

We can consider these questions in order.

Topic Modeling: Topic Segmentation and Topic Labeling Conversations often span different topics; an initial email message, asking a team to explain low sales in Asia, can generate a thread on what is the best visualization tool for a particular analysis task. Or, alternatively, the follow up may be a discussion on how the team may need to be reorganized.

Even if you look at our short, sample email conversation, it clearly covers at least two topics. The conversation starts with a proposal for a vacation but then one sub-thread (on the right of Figure 1.4) veers off talking about a problematic course assignment.

This example can help us to define the two basic subtasks of topic modeling: topic segmentation and topic labeling. In topic segmentation, you are interested in identifying what portions of the conversation are about the same topic, or equivalently, in detecting where in the conversation the topic shifts are. For instance, in our sample conversation, there is a topic shift between the first and the second (non quoted) sentences in Email-1.2 and this shift splits the conversation in two segments, i.e., the text below the shift in the right sub-thread vs. the rest of the conversation.

Topic labeling, on the other hand, is about generating informative labels (typically sets of words) for all the topics covered by a conversation. In our example, two informative (but still not ideal) labels for the two identified topics might be "spring break Mexico skiing" and "assignment question idea".

A large number of topic modeling techniques have been developed for generic text (not necessarily conversational in nature), including supervised and unsupervised machine learning methods, as well as a combination of the two. Among all these proposals, a novel, probabilistic approach, based on Latent Dirichlet Allocation (LDA) [Blei et al., 2003] appears to be the most effective and influential (see Blei and Lafferty [2009] for a gentle introduction). In LDA, the generation of a collection of documents is modeled as a stochastic process, and topic modeling consist of estimating the parameters of the underlying probabilistic generative model.

In Chapter 3, we will discuss how topic modeling techniques developed for generic text can be extended to deal with text conversations. For instance, how variations of the LDA framework have been successfully applied to meeting transcripts [Purver et al., 2006b], as well as to Twitter [Ramage et al., 2010] and email conversations [Dredze et al., 2008].

Sentiment and Subjectivity (i.e., Opinion Mining) Conversations typically exhibit a large amount of highly subjective content. Participants may agree or disagree with one another, argue for or against various proposals, and generally take turns expressing their opinions and emotions. Mining all this subjective content can be framed at different levels of granularity. At the highest level, you have the task of detecting whether a contribution to the conversation is subjective or not. This can be further refined by determining whether a contribution that has been identified to be subjective expresses a positive vs. a negative sentiment. Finally, at an even more specific level, you may want to determine the strength of the expressed opinion, for instance in the interval [-3, +3], with a zero value for non-subjective (i.e., neutral) contributions. For illustration, if the sentence *"I applaud the new budget proposal."* was considered, it should be classified as expressing a positive opinion, with strength +3. And if we look at our sample email conversation, sentences like *"great idea!"*, *"I do not like this assignments"*, *"Let's talk tomorrow at UBC after class"* should be, respectively, classified as positive (+3), negative (-2) and neutral (0).

Numerous techniques have been proposed in the literature to perform opinion mining from generic text [Pang and Lee, 2008]. Most of these techniques rely on lexical and syntactic features of the text, meaning that they look for the presence (or absence) in the target text of particular words, as well as of particular syntactic constructions. For instance, two features commonly used are the presence of adjectives with a positive or negative orientation (e.g., *interesting* vs. *boring*) and the presence of syntactic patterns involving negation like $not < intensifier >< adjective >$, which would match the phrase *not very inspiring*. Notice that machine learning methods are often used, in order to determine what features are most useful for opinion mining, out of the large number of candidates (e.g., [Wilson et al., 2006]).

In Chapter 3, we will see how approaches to opinion mining, based on lexical and syntactic features, can be directly applied to text conversations. That chapter will also discuss how the set of features can be expanded to include conversational features. For instance, features related to the speaker of the particular contribution, as well as the position of the contribution within the conversation.

Extracting the Conversational Structure Conversations have properties that clearly distinguish them from monologues. To have a conversation, you need two or more participants who exchange information by talking or in writing. Each contribution to a conversation, technically each turn, is performing one or more dialogue acts; for instance, making a statement, asking or answering a question, and making or accepting a request. These dialogue acts tend to occur in pairs, called adjacency pairs , where the first turn from one participant is generating the following turn by another participant. Common examples of these pairs are, for example, a question followed by an answer and a request followed by and acceptance or a rejection. For illustration, in our sample email conversation the request expressed in Email-1 *"Would you like to join me?"* is followed by an uncertain response in Email-1.1, and by reject response in Email-1.2.

dialogue acts

adjacency pairs

The structure of a conversation specifies how all the dialogue acts that comprises the conversation are connected to each other. In synchronous conversation like spoken ones and instant messaging, the structure of the conversation can be expected by and large to be linear, as turns must occur one after the other, with minimal delay and minimal overlap. In contrast, asynchronous written conversations, like emails and blogs, often display a more complex structure, as consecutive turns can be far apart in time, multiple turns can largely overlap, and turns tend to express more than one dialogue act each.

Mining the conversational structure of a conversation involves at least the following sub-tasks: recognizing what dialogue act(s) are performed by each turn in the conversation, connecting them to form adjacency pairs, and reconstructing the possibly complex structure of the conversation.

In Chapter 3, we will see how this tasks can be performed, with different degree of success, by applying machine learning techniques, which include supervised, semi-supervised and unsupervised ones. In Chapter 3, we will also discuss mining tasks that can be useful, when the focus of the conversation is to make a joint decision and come up with a set of action items.

Evaluation of Mining Tasks Many of the tasks involved in mining text conversation can be effectively framed as classification problems, in which for instance a sentence (or a turn) is classified as performing a particular dialogue act, or as expressing a positive vs. negative opinion. Standard evaluation metrics for classification, including Precision, Recall, F-score and AUROC, are described in Chapter 2.

1.4.2 SUMMARIZING TEXT CONVERSATIONS

Once a text conversation is mined for topics, opinions and conversational structure, the next challenging question is: How can we summarize the conversation? Or, in other words, how can we identify what are the most salient parts of the conversation and effectively present those to the user?

Research on text summarization has a long history in NLP [Jurafsky and Martin, 2008]. Subareas of the field can be identified by looking at what the input to the process is, the way in which the summary is generated, how the summary is presented to the user, and the intended function of the generated summary.

In this section, we clarify the fundamental dimensions for summarization research and examine how they relate to the problem of summarizing text conversation. We will also briefly discuss metrics for evaluating the effectiveness of a summarizer.

Single vs. Multi-Document Input With respect to the input, two key tasks have been considered. *Single-document summarization*, where the goal is to summarize a single document (e.g., a news article) vs. *multi-document summarization*, in which the information necessary to satisfy the user information needs is distributed across different documents (e.g., all the new articles about a particular event). In summarizing text conversation this distinction is relevant, for instance, to differentiate between summarizing a single contribution (e.g., email [Muresan et al., 2001]) vs. summarizing a whole conversation (e.g., email thread [Rambow et al., 2004]), which can be seen as a group of distinct but interconnected documents, one for each message/contribution. In Chapter 4, we will see that the single and multi document summarization tasks are not that orthogonal to each other in summarizing text conversations. Hybrid approaches can been devised in which the summarization of a message is performed and presented in the context of the preceding conversation (e.g., [Lam et al., 2002]). And a summary of the preceding conversation could arguably provide the most suitable

single-document
multi-document

context. A simple application of this approach is shown in Figure 1.5, where Email 1.2.1 from our sample conversation is summarized in the context of a relevant contribution from the preceding conversation.

Previous Relevant Conversation

Kai
...Do you guys have any idea on how to deal with Q1?

Summary of Email-1.2.1

Mark (reply-all)
....
I am also having problems.

Figure 1.5: A single document summary for Email-1.2.1 in our synthetic email conversation. The summary is simply an informative sentence extracted from that email and context is provided from the preceding conversation.

Extractive vs. Abstractive Methods We also have a binary distinction in the way a summary can **extractive** be generated. In *extractive summarization* a summary is produced by simply selecting a subset of the sentences of the input documents. In practice, a measure of informativeness is computed for each sentence (often by a ML classifier) and the most informative sentences are selected and ordered. Notice that reordering the selected sentences is only a problem for extractive multi-document summarization, as for extractive single-document summarization sentences can appear in the summary in the same order as they appeared in the source document. The extractive approach has been by far the most popular in the literature, largely because extractive systems do not require the generation of novel language, since the summary sentences are simply lifted from the source document(s). The extracted sentences may then be compressed, simplified or aggregated in some way.

abstractive An alternative approach to generate summaries is by *abstractive summarization*, which tries to reflect more closely what people naturally do when they summarize text. In an abstractive summarizer, knowledge is first extracted from the source documents. Next, new knowledge is derived by inference or by aggregating and abstracting knowledge (that could have been extracted from multiple sentences). Finally, the most informative content is selected and expressed in natural language. Abstractive summarization is arguably a much more complex and challenging process than extraction, since it requires not only a natural language generation module, but often also a domain dependent component to process and rank the extracted knowledge.

For illustration, Figure 1.6 shows two abstractive summaries of our sample email conversation, while Figure 1.7 shows one extractive summary of the same conversation. Notice how the level of abstraction in abstractive summarization can vary considerably, with the first abstractive summary being much more abstract than the second one.

(i) Friends discuss vacation plans and an assignment

(ii) Jane propose to Mark and Kay to go to Mexico. Neither can go.
Jane then propose to go skiing but this idea is also abandoned.
Kai and Mark have problems with an assignment. Jane will help them.

Figure 1.6: Two abstractive summaries of our synthetic email conversation.

In Chapter 4, we will see that most of the summarizers for text conversations developed so far are fundamentally extractive in nature. However, in that chapter, we will also cover a few very recent studies on applying abstractive summarization to text conversations [Murray et al., 2010].

Generic vs. Query-based Summarization Another important dimension related to the input of the summarization process is whether the user is explicitly stating her information needs by means of a query. If this is the case, a good summary should not be generated generically, but should focus on the query, which, for instance, could refer to a particular event, date or person. In practice, a query-based summarizer can focus on the query by taking the query into account when deciding whether to include some content (a sentence or a piece of information) in the summary. This is typically done by measuring the overlap/similarity between that content and the query. A similar approach can be followed for text conversations. For instance, a common feature used for measuring informativeness in email summarization is subject-line overlap or similarity (e.g. [Nenkova and Bagga, 2003]). If we combine the subject line with a user-provided query, we can generate query-dependent summaries that tailor the summary to a particular information need. As another example, consider work by Sharifi et al. [2010], where the task is automatically summarizing microblogs such as Twitter messages. The algorithm takes as input a topic phrase (e.g., Ice Dancing) along with a set of sentences from relevant tweets and it generates an extractive, query-based summary intended to concisely convey why the topic is currently popular on Twitter (e.g., "'Ice Dancing Canadians Tessa Virtue and Scott Moir clinch the gold in Olympic ice dancing; U.S. pair Davis and White win silver; 2/22/2010'").

query-based

For an example of a query-based abstractive summary of our synthetic email conversation, see Figure 1.8.

Jane:
I was thinking of going to Mexico over the spring break.

Mark
great idea!

Jane (reply-all)

What about going skiing for a few days?

Mark (to Jane)
see you in class or after the break

==========

Kai
No Mexico, sorry, I need to catch up with 502.
...Do you guys have any idea on how to deal with Q1?

==========

Mark (reply-all)

....
I am also having problems.

==========

Jane
Let's talk tomorrow at UBC after class.

Figure 1.7: Sample extractive summary of our synthetic email conversation.

Indicative vs. Informative Summarization We can discriminate summaries based on how they are intended to function with regards to the source document(s). *Informative* summaries attempt to convey the most important information of a document. The notion of substitutability is central to the idea of an informative summary, as the summary should convey enough of the critical information that it is able to stand in for the source document. On the other hand, *indicative summaries* give a high-level outline of the document but do not attempt to convey all of the critical information from the source. They are typically provided so that the reader can decide whether or not to start reading the source document. In practice, a given summary may be a mixture of both types. As one

informative

indicative

QUERY= Mexico

Summary
Jane propose to Mark and Kay to go to Mexico. Neither can go.

Figure 1.8: Sample query-focused (abstractive) summary of our synthetic email conversation.

example, a human-authored abstract in a science journal will usually give a high-level overview of the experiments and conclusions, but may highlight a key finding in some detail.

Domain-Specific vs. General-Purpose Summarization We have mentioned conversation types such as emails, meetings, blogs and chats, and we refer to these as separate conversation *modalities*. Modality here refers to a means or mode of communication, where a particular conversation modality may **modality** be associated with both distinct communication technologies as well as distinct social conventions and language characteristics. From a more general viewpoint, without reference to communication or language, these can also be considered distinct *domains*, and we will use the two terms more or **domain** less interchangeably here.

For many tasks there is a tension between developing solutions that are general and broadly applicable, and implementing tools that work only in specific domains, but are highly effective. Summarization is not an exception in this respect. Researchers have worked both on domain specific systems (e.g., McKeown et al. [2002]) for news, Zhou et al. [2004] for biographies) and on general purpose platforms [Radev et al., 2004]. A related distinction for summarizing text conversation, that will be discussed in Chapter 4, is whether a summarization approach can be only applied to a particular conversational modality (e.g., email), or whether it can work on any text conversation, independently from its modality. While most of the summarizers described in Chapter 4 are domain/modality specific, as they exploit peculiar features of those modalities (e.g., the subject line for emails, user ratings for blog posts), we will also cover recent attempts to design a multi-modal system [Murray and Carenini, 2008] that relies only on features common to all multi-party interaction, such as speaker dominance in the conversations, turn-taking, lexical cohesion, etc. This system is not only capable of summarizing conversations in different modalities (e.g., meeting, emails, blogs), but it can also work on conversations spanning multiple modalities (e.g., a transcript of a meeting that was followed up by an email conversation). A multi-modal approach presents two additional, critical advantages. First, by only harnessing features shared by all the modalities, it can facilitate the transfer of knowledge from one modality to another [Sandu et al., 2010], which in machine learning is called domain adaptation [Daumé and Marcu, 2006]. Secondly, this general approach should **domain adaptation** easily cover novel conversational modalities that are being constantly created by people's creativity and technological advancements.

WORD CLOUD VISUALIZATION

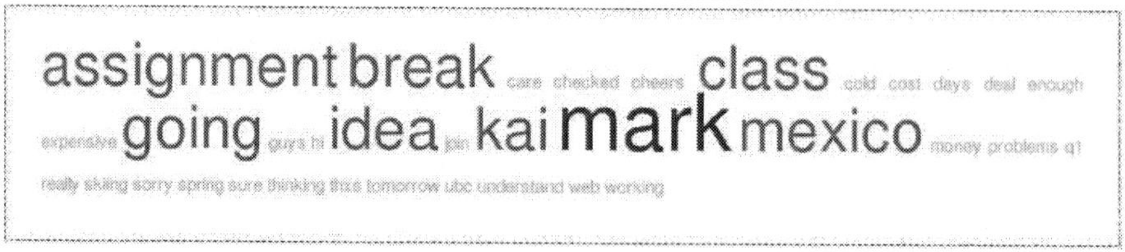

TEXTUAL SUMMARY

Friends discuss vacation plans and an assignment

Figure 1.9: Simple multimedia Summary of our synthetic email conversation.

Textual vs. Multimedia Output Traditionally, research on text summarization has been about taking as input one or more documents and generating as output a textual summary of those documents. As we have already seen, even if we are restricted to textual output, there are two possibilities. The output can be either a subset of the sentences from the input (i.e., an extract), or a set of novel sentences that are automatically generated to describe the most important content extracted from the input (i.e., an abstract).

Depending on the user task and information needs, more possibilities can be envisioned, if we move beyond textual summaries. Arguably, all the information mined from a conversation could be conveyed graphically. For instance, extracted topics could be visualized like a *theme river*, in which the temporal evolution of the strength of different topics is depicted as a multi-colored visual river flowing from left to right [Havre et al., 2002]. Similarly, extracted opinions can be also effectively conveyed graphically. Pang and Lee [2008] present some illustrative examples in Chapter 5 of their book (Summarization Chapter). More generally, any information visualization for text analysis, like Word Trees and Word Cloud, could be effectively applied to text conversations [Hearst, 2009](Chapter 11).

information visualization

It is widely known that text and graphics are not mutually exclusive, but can actually complement each other. For instance, Carenini et al. [2006] present a multimedia opinion summarization system, in which a visualization of the extracted opinions is integrated with a textual summary, to support the user in the interactive exploration of the source documents. In Chapter 4, we will see that similar approaches can be applied to text conversations.

A simple example of a multimedia summary of our sample email conversation is shown in Figure 1.9.

Summarization evaluation As with all mining and retrieval tasks, it is critical to have dependable summarization evaluation metrics to assess various systems. It is also important to have *widely used* evaluation schemes so that researchers can compare results directly with one another and determine the state of the art. In recent years, several approaches to evaluation have become popular within the summarization community and adopted for periodic benchmark tasks.

Intrinsic evaluations measure the information content of a generated summary, typically by comparing it with human gold-standard summaries. These types of evaluations are concerned with whether the candidate summary contains the most important information from the source document. Many of the intrinsic evaluation schemes we will introduce are automated metrics, and as such it is important to confirm that they correlate with human judgments. A major reason why the summarization community has been slow to adopt "official" evaluation metrics (compared with, say, the machine translation community) is precisely owing to conflicting results regarding such correlations in different domains. Liu and Liu [2010] is a recent example of work trying to measure the usefulness of a popular intrinsic evaluation software package (ROUGE, described in Chapter 2) on noisy conversational data.

intrinsic evaluation

Extrinsic evaluations, on the other hand, measure the usefulness of a summary in aiding some real-world task, such as document classification or reading comprehension. The motivation for conducting extrinsic evaluations is that summaries are generated for some purpose, and we should directly evaluate how well they serve that purpose, rather than simply comparing them with other summaries. However, extrinsic evaluations are typically user studies, which involve a great deal of human hours in terms of design, recruitment, experiments and analysis. It is therefore common to regularly employ intrinsic evaluations to speed research and development, while occasionally carrying out extrinsic evaluations to assess major development milestones.

extrinsic evaluation

1.5 BOOK PREVIEW

In Chapter 2, we describe popular conversation corpora for summarization and mining research, including descriptions of the relevant annotations. We also describe in detail the widely used evaluation metrics for both text mining generally and automatic summarization particularly.

In Chapter 3, we introduce mining tasks and methods for conversational data. This includes topic segmentation and labeling, subjectivity and sentiment detection, dialogue act detection, extraction of conversation structure, and detection of decisions and action items.

In Chapter 4, we first give a general characterization of the architecture of summarization systems, then describe how summarizers have been designed for particular conversation modalities. We also describe attempts at developing summarizers for conversations across modalities, and give a detailed case study of an abstractive, multi-modal conversation summarizer.

In Chapter 5, we review our discussion and lay out suggestions for future work in the promising and still largely unexplored corners of the mining and summarization research space.

Assumptions about Our Readers We have tried to make this book accessible by providing sufficient background on each topic, and think that it should be suitable for the graduate student who may have a background in computer science or linguistics but only minimal exposure to NLP. However, due to space limitations, we do assume that our readers are at least somewhat familiar with several topics, including basic probability and machine learning. In Section 1.3, we have provided pointers to key background references in NLP, AI and machine learning. Furthermore, in each section we have supplied pointers to further reading, including entry-level primers on most topics.

Conventions and Notations Throughout the book, we have highlighted key terms in the margin when they are first introduced, for ease of reference. At the conclusion of Chapters 2, 3 and 4, we summarize the important points of the chapter and give suggestions for further reading.

CHAPTER 2

Background: Corpora and Evaluation Methods

In this chapter we describe some of the conversation datasets that are widely used for summarization and text mining research. Large collections of possibly annotated documents are called *corpora* (sing. *corpus*) in NLP and we will use this terminology. We characterize the raw data as well as the available annotations. Most of the techniques presented in this book rely on machine learning methods that need to be trained and tested using such corpora. Subsequently, we detail the evaluation metrics that are commonly used for summarization and text mining tasks.

2.1 CORPORA AND ANNOTATIONS

In this section, we introduce two meeting corpora and two email corpora, all of which are freely available. We describe the annotations (or *codings*) that are most relevant and useful for summarization and text mining. When we say that a corpus has been annotated or coded for a particular task such as summarization, we mean that human judges have manually labeled the data for the phenomena relevant to that task. For summarization, this typically means identifying the most important sentences and writing a high-level abstract summary of the document, but we will describe such annotation schemes in detail momentarily.

annotation

At points we refer to the κ statistic for a given set of annotations, which measures agreement between multiple annotators, factoring in the probability of chance agreement [Carletta, 1996]. More precisely, κ is used to measure agreement between each pair of annotators where the annotators are making category judgments. In the case of extractive summarization, for example, the category judgment is whether or not each sentence should be extracted. In the case of opinion mining, to make another example, the judgment is whether the sentence has a positive, negative or neutral polarity.

kappa statistic

Given two sets of codings representing the category judgments of two annotators, κ is calculated as

$$\kappa = \frac{P(A) - P(E)}{1 - P(E)},$$

where P(A) is the proportion of times the annotators agree with one another and P(E) is the proportion of agreement that we would *expect* based purely on chance. When multiple coders are carrying out annotations on the same data, we expect some baseline level of agreement just by chance.

chance agreement

We want our κ statistic to tell us whether actual agreement is above that baseline level. If not, κ is zero. At the other extreme, perfect agreement would yield a κ statistic of one.

As we will see, it is not uncommon to have very low κ scores for a given corpus, particularly with the summarization task. The fact that the κ scores for summarization annotation are well below one does not mean they are useless, but merely that there is no such thing as a "single best summary." And for that reason, it is important to recruit as many annotators as possible, and as many annotations per document as possible, when doing summarization coding.

2.1.1 MEETING CORPORA

Meetings represent one of the conversational domains that has received the most attention from the mining and summarization communities. Research on meetings has been greatly facilitated in recent years by the availability of large, freely available annotated corpora. We discuss two meeting corpora in particular, the AMI corpus and the ICSI corpus, and describe the manner in which they were annotated for summarization purposes and for a variety of other mining tasks.

AMI Corpus The AMI meeting corpus [Carletta, 2006] was created as part of the European Union-funded AMI project[1]. The corpus consists of ~100 hours of recorded, transcribed and annotated meetings, divided into *scenario* and *non-scenario* meetings. In the scenario meetings, four participants take part in each meeting and play roles within a fictional company. The scenario given to them is that they are part of a company called Real Reactions, which designs remote controls. Their assignment is to design and market a new remote control, and the members play the roles of project manager (the meeting leader), industrial designer, user-interface designer, and marketing expert. Through a series of four meetings, the team must bring the product from inception to market.

The first meeting of each series is the kick-off meeting, where participants introduce themselves and become acquainted with the task. The second meeting is the functional design meeting, in which the team discusses the user requirements and determines the functionality and working design of the remote. The third meeting is the conceptual design of the remote, wherein the team determines the conceptual specification, the user interface, and the materials to be used. In the fourth and final meeting, the team determines the detailed design and evaluate their result.

The participants are given real-time information from the company during the meetings, such as information about user preferences and design studies, as well as updates about the time remaining in each meeting. While the scenario given to them is artificial, the speech and the actions are completely spontaneous and natural. There are 138 meetings of this type in total. The length of an individual meeting ranges from ~15–45 minutes, depending on which meeting in the series it is and how quickly the group is working.

The non-scenario meetings are naturally occurring meetings that would have been held regardless of the AMI data collection, and so the meetings feature a variety of topics discussed and a variable number of participants.

[1] `http://www.amiproject.org/`

The meetings were recorded at three European locations. The participants consist of both native and non-native English speakers, and many of them are students.

The AMI corpus is freely available[2] and contains numerous annotations, such as the summarization annotation described below, and multi-modal artefacts such as PowerPoint slides, notes, and whiteboard events.

ICSI Corpus The ICSI meeting corpus [Janin et al., 2003] is a corpus of 75 natural (i.e., non-scenario) meetings. As with the AMI non-scenario set, these are meetings that would have been held anyway and feature a variable number of participants. Because many of the meetings in the corpus are gatherings of ICSI researchers themselves, the topics tend to be specialized and technical, e.g., discussions of speech and language technology. The average length of an ICSI meeting is approximately one hour, which is greater than the average AMI non-scenario meeting (~15-45 minutes).

Like the AMI corpus, the ICSI corpus meetings feature both native and non-native English speakers. All meetings in the corpus were recorded at ICSI in Berkeley, California. Unlike the AMI scenario meetings and similar to the AMI non-scenario meetings, there are varying numbers of participants across meetings in the ICSI corpus, with an average of six but sometimes as many as ten per meeting.

Unlike the AMI corpus, which is multi-modal and contains a variety of information such as slides, whiteboard events and participant notes, the ICSI corpus consists entirely of speech and relevant annotations. The ICSI corpus can be freely downloaded[3] and additional annotations of the ICSI meetings are available via the AMI corpus download. Both corpora were annotated with similar summarization annotation schemes as part of the AMI project, and we will describe those annotations shortly. However, we first describe some basic concepts necessary to understand these annotations.

Utterances, Dialogue Acts and Disfluencies Usually when we talk about extractive summarization, we are talking about extracting sentences from a document. However, with spoken conversations such as meetings, people typically do not speak in complete, well-formed sentences. Their *utterances* may be disfluent and ungrammatical. The utterances may be peppered with *filled pauses* such as *uh* and *um*, indicating that the speaker is thinking. Utterances may overlap as speakers interrupt one another, or a sentence may be abandoned if the speaker thinks the listener already understands. Filled pauses, repetitions and fragments are all examples of *disfluencies*, phenomena which tend to make speech less fluent and grammatical. Disfluencies can particularly pose a problem when transcribing speech, as the resulting transcript can be difficult to read if the disfluencies are not corrected or removed.

utterances

disfluencies

As we saw in Section 1.4.1, one way utterances can be analyzed is by identifying *dialogue acts* [Stolcke et al., 2000]. A dialogue act represents the illocutionary meaning of an utterance, or

dialogue act

[2] `http://corpus.amiproject.org/`
[3] `http://www.idiap.ch/mmm/corpora/icsi`

the action performed by the utterance. For example, the utterance *Is that right?* has the dialogue act type *Backchannel-Question* because it is a question and it is also steering the conversation back to the previous speaker. The most frequent type of dialogue act in meeting corpora is *Statement*, where a speaker is simply providing information. So utterances can be classified as dialogue acts, where each is assigned its most likely dialogue act *type*.

In the discussion that follows, we often use *utterance* and *dialogue act* interchangeably, where we are referring to a sentence-like unit in speech. In other cases we will refer to specific dialogue act types such as *Yes-No-Question*, *Agreement* or *Statement*.

AMI and ICSI Meeting Annotation Both the AMI and ICSI corpora were initially annotated for topic segmentation (see Section 1.4)and dialogue acts. Note that the MRDA (Meeting Recording Dialogue Act) corpus refers to dialogue act annotations of the ICSI meetings. These annotations are described in detail by Shriberg et al. [2004]. The two meeting corpora were then annotated for summarization. More specifically, for the summarization annotation, annotators were asked to write abstractive summaries of each meeting and to extract the meeting dialogue acts that best convey or support the information in the abstractive summary. As described in Chapter 1, an abstractive summary is a high-level summary using novel text to synthesize and describe information from the document.

Annotators used a graphical user interface (GUI) to browse each meeting, enabling them to view previous human annotations consisting of a written transcription synchronized to the meeting audio, and topic segmentation. The annotators were first asked to build a textual summary of the meeting aimed at an interested third-party, using four sub-headings for their abstract. For the ICSI meetings, those four headings are:

- general abstract: "why are they meeting and what do they talk about?";
- decisions made by the group;
- progress and achievements;
- problems described

For the AMI meetings, the summary sections vary slightly:

- general abstract;
- decisions;
- actions;
- problems.

The maximum length for each section of the abstract is 200 words, and while it was mandatory that each general abstract section contained text, it was permitted that for some meetings the other three sections could be empty; for example, some meetings might not involve any decisions being made. Annotators were encouraged to listen to a meeting in its entirety before beginning to compose the summary.

After authoring the abstractive summary, annotators were subsequently asked to create an extractive summary, using a second GUI. As described in Chapter 1, an extractive summary is comprised of sentences, or, in this case, dialogue acts, taken from the document. With this GUI they were able to view their abstract summary and the transcript of the meeting, with the topic segments removed and with one dialogue act per line. They were given the pre-existing dialogue act coding [Shriberg et al., 2004] and viewed only the dialogue act segments without the dialogue act type labels. They were instructed to extract the dialogue acts that taken together could best convey the information in the abstractive summary and could be used to support the correctness of that abstract. They were not given any specific instructions or limitations about the number or percentage of dialogue acts to extract, nor any instructions about extracting redundant dialogue acts. They were then required to do a second pass of annotations, wherein for each extracted dialogue act they chose the abstract sentences supported by that dialogue act. The result is a many-to-many mapping between abstract sentences and extracted dialogue acts; that is, an abstract sentence can be linked to more than one dialogue act, and vice versa. Although the expectation was that each abstract sentence would be linked to at least one extracted dialogue act and each extracted dialogue act linked to at least one abstract sentence, annotators were allowed to leave abstract sentences and dialogue acts standing alone in some circumstances.

In addition to the annotation just described, Wilson [2008] annotated 20 AMI meetings for a variety of subjective phenomena at the dialogue act level which fall into the broad classes of *subjective utterances*, *objective polar utterances* and *subjective questions*. Two subclasses of subjective utterances are *positive subjective* and *negative subjective* utterances. Such subjective utterances involve the expression of a private state [Quirk et al., 1985] (an emotion or state of mind that is not always observable), such as a positive/negative opinion, positive/negative argument, and agreement/disagreement. An objective polar utterance is one that conveys positive or negative facts without indicating any private state, e.g., "*The camera broke the first time I used it*" is a negative polar utterance [Wilson, 2008]. Subjective questions are when a speaker enquires about the opinions or feelings of another person. The 20 meetings were labeled by a single annotator, although Wilson [2008] did conduct a study of annotator agreement on two meetings, reporting a κ of 0.56 for detecting subjective utterances.

2.1.2 EMAIL CORPORA

While emails have been a popular domain for mining and summarization work, a potential research bottleneck is finding a sufficient amount of available email data. Research groups may have their own email corpora, but privacy concerns often preclude their general release. However, in recent years, two email corpora have been made publicly available and numerous researchers are now often working on the same datasets, making their results easier to compare.

Enron Corpus The Enron email corpus is a collection of emails released as part of the investigation into the Enron corporation [Klimt and Yang, 2004]. It has become a popular corpus for NLP research (e.g. [Bekkerman et al., 2004, Chapanond et al., 2005, Diesner et al., 2005, Yeh and Harnly, 2006]) due to being realistic, naturally occurring data from a corporate environment, and moreover because, as previously mentioned, privacy concerns mean that there is very low availability for other publicly available email data. The email dataset is freely available to download[4]. The released dataset features approximately half a million email messages and about 150 people.

At the University of British Columbia, 39 threads of the Enron corpus have been annotated for extractive summarization, with five annotators assigned to each thread [Carenini et al., 2007]. The annotators were asked to select 30% of the sentences in a thread, subsequently labeling each selected sentence as either "essential" or "optional." Essential sentences are weighted three times as highly as optional sentences. A sentence score can therefore range between 0 and 15, with the maximum score achieved when all five annotators consider the sentence essential, and a score of 0 achieved when no annotator selects the given sentence. The Enron corpus summary annotations are available upon request.[5]

BC3 Corpus The BC3 corpus [Ulrich et al., 2008] contains email threads from the World Wide Web Consortium (W3C) mailing list. The threads feature a variety of topics such as Web accessibility and planning face-to-face meetings. The annotated portion of the mailing list consists of 40 threads and 261 individual emails. The threads are annotated in a similar manner as the AMI corpus, with three human annotators per thread first authoring abstracts and then linking email thread sentences to the abstract sentences. The corpus also contains annotations indicating requests, commitments, questions, and other dialog acts. It can be freely downloaded.[6]

The full W3C dataset—used in the TREC Enterprise Track[7]—includes mailing lists, public webpages, and text derived from .pdf, .doc and .ppt files, among other types. However, only the mailing list data was used for the BC3 corpus. The mailing list subset is comprised of nearly 200,000 documents, and TREC participants have provided thread structure based on reply-to relations and subject overlap.

As we said before, in the BC3 corpus, the W3C emails have been annotated for summarization as well as labeled for several sentence-level linguistic phenomena. The annotation procedure went as follows. After familiarizing themselves with a given email thread, annotators were asked to write an abstractive summary of the thread. Then, for each abstractive sentence, they were instructed to list the email sentence IDs that correspond to the abstractive sentence. This results in a many-to-many mapping between extract sentences and abstract sentences for each annotator. The judges were told that under some circumstances an abstractive sentence could remain unlinked to any extract sentence, but that this should be a rare occurrence.

[4]`http://www.cs.cmu.edu/~enron/`
[5]`http://www.cs.ubc.ca/nest/lci/bc3.html`
[6]`http://www.cs.ubc.ca/nest/lci/bc3.html`
[7]`http://trec.nist.gov/data/enterprise.html`

After the creation of the abstract and the relevant linking, annotators were allowed to select email sentences which they considered important but were not linked to the abstract. Likewise, they could remove a linked email sentence from their extract if it was considered unimportant despite being linked to the abstract. This annotation scheme allows researchers to closely investigate the relationship between extracts and abstracts. The scheme closely follows the methods used by researchers in the AMI project in annotating their meeting corpus [Carletta, 2006].

Three people annotated each thread. Their annotations had a κ agreement of 0.50 for the extracted sentences. This compares to a κ statistic of 0.45 in the AMI corpus [Carletta, 2006] for meeting summarization, and 0.31 in the ICSI corpus [Janin et al., 2003] for meeting summarization. A total of 10 recruits were used for the annotation.

Annotators were also asked to label a variety of sentence-level phenomena, including whether each sentence was subjective. In a second round of annotations, three different annotators were asked to go through all of the sentences previously labeled as subjective and indicate whether each sentence was *positive*, *negative*, *positive-negative*, or *other*. The definitions for positive and negative subjectivity mirrored those given by Wilson [2008] and used for annotating the AMI corpus, mentioned above.

2.1.3 BLOG CORPORA

To our knowledge, there is not a freely available corpus of *conversational* blog data complete with annotations for summarization and mining purposes. Perhaps the most widely used blog corpus for automatic summarization research is the dataset released as part of the Text Analysis Conference (TAC, formerly known as the Document Understanding Conference, or DUC) 2008 track on opinion summarization[8]. This dataset consists of blog posts on a variety of given topics. The task was to automatically summarize opinions on a person, entity or topic by analyzing numerous blog posts on that topic. For example, one cluster of blog posts related to the company *Jiffy Lube* and the task was to summarize what people think of that company. However, the blog posts are not truly conversational; individual posts do not include comments and the posts do not link or refer to each other.

We believe it would be of great benefit to the research community to annotate and release a corpus of blog conversations. This entails more clearly defining summarization and mining tasks for blog data. In some cases, we may be interested in analyzing how a set of blog comments reflects on, or expands upon, the initial post. In other cases, we may want to analyze blog conversations much more widely, by analyzing how bloggers link and respond to one another across blogs.

2.2 EVALUATION METRICS FOR TEXT MINING

In this section we discuss evaluation metrics that are commonly used for a wide variety of text mining tasks such as summarization, sentiment detection and topic modeling.

[8] `http://www.nist.gov/tac/2008/summarization/`

2.2.1 PRECISION, RECALL AND F-SCORES

supervised

Many of the mining and summarization techniques described in this book are supervised binary classifiers, where *supervsied* means that the classifier requires training on labeled data and *binary* means we are predicting one of two classes. For example, a classifier that discriminates subjective from non-subjective comments or informative from non-informative sentences may be trained on data where each sentence has been labeled as belonging to one of the two classes. In other words, with all of these tasks we are trying to discern a positive class from a negative class. In these cases, we can evaluate the classifier using precision, recall and F-score. Precision and recall are calculated as follows:

precision and recall

$$Precision = TP/(TP + FP)$$

$$Recall = TP/(TP + FN) \ ,$$

where TP means true positives (correctly classified as positive), FP means false positives (incorrectly classified as positive), and FN means false negatives (incorrectly classified as negative). Note that these two measurements share the same numerator, TP, the number of items correctly classified as positive. To get precision, we divide the numerator by the number of items that were *predicted* to be positive. To get recall, we divide the numerator by the number of items that really *are* positive. A perfect classifier would have both precision and recall equal to 1, as FP and FN would be equal to 0 (i.e., no data would be incorrectly classified as positive or negative).

F-score

The F-score is simply a combination of precision and recall. The harmonic mean is typically used, which is given by the following equation when precision and recall are weighted equally:

$$F = \frac{2 * Precision * Recall}{Precision + Recall} \ .$$

The perfect classifier would have an F-score equal to 1.

2.2.2 ROC CURVES

posterior probability

Many of the mining and summarization techniques described in this book rely on probabilistic binary classifiers, which assign to each data instance (e.g., a sentence) a *posterior probability* of belonging to a certain class, given the evidence, e.g., the sentence features used.

When calculating precision, recall and F-score for a probabilistic classifier, we evaluate the classifier at a particular posterior probability threshold, where we consider a data instance to be "positive", i.e., to belong to the class, if the classifier's posterior probability for that particular instance is greater than or equal to a threshold and "negative" otherwise. A commonly used threshold is 0.5, the mid point of the [0, 1] probability range.

Arguably, a more informative alternative is to evaluate the classifier across all possible probability thresholds between 0 and 1. In practice, we can measure the true-positive/false-positive rates as the posterior threshold is varied. The true-positive rate (TPR) and false-positive rate (FPR) are calculated as follows:

$$TPR = TP/(TP + FN)$$

$$FPR = FP/(FP + TN) .$$

Now, for example, if we set the probability threshold at 0 for our classifier, then all instances are considered positive (all have probabilities equal or greater than 0), giving a true-positive rate (TPR) of 1 and a false-positive rate (FPR) of 1. The reason for this is that, since nothing is classified as negative, we have that $FN = TN = 0$.

At the other extreme, if we set the threshold at 1 for our classifier, then all instances are considered negative (none has probability equal or greater than 1)[9], giving a true-positive rate (TPR) of 0 and a false-positive rate (FPR) of 0. The reason for this is that, since nothing is classified as positive, we have that $TP = FP = 0$.

By varying the posterior threshold by small decrements from 1 to 0, measuring the TPR and FPR at each stage, we end up with a receiver operator characteristic (ROC) curve . This curve is created by plotting all the TPR and FPR values in a Cartesian quadrant with FPR on the [0,1] x-axis and TRP on the [0,1] y-axis (see Figure 2.1). As we said, when we start with the threshold equal to 1, we have a point in the origin, while when we end with a threshold equal to 0, we have a point in the upper right corner. **ROC curve**

A curve that is close to a diagonal line would indicate random performance and a poor classifier, because as we vary the threshold form 1 to 0, the TPR increases at the same rate as its FPR (Figure 2.2).

A better classifier is represented by Figure 2.3, where the TPR increases faster than the FPR. A good classifier would yield a ROC curve that rises steeply vertical along the lefthand side of the graph before moving horizontally to the right along the top of the graph. Such a classifier is shown in Figure 2.4. While we label this a "good" classifier, that is only in relation to the others, and in fact for some critical tasks (such as medical testing) this performance may not be nearly good enough. Different tasks have different tolerances for the proportion of positive instances labeled as negative (i.e. false negatives) and negative instances labeled as positive (false positives). Note also that a ROC curve that is well *below* the diagonal would also indicate a good classifier, albeit one which has its class labels swapped.

Because the ROC is simply a visual indicator and not a numeric measurement, it is customary to also calculate the area under the ROC curve (AUROC) by dividing that area into trapezoidal spaces. This yields a number between 0.5 (diagonal line and random performance) and 1 (perfect performance). It is easy to tell at first glance that the classifier represented by Figure 2.2 has an AUROC of 0.5 since it clearly divides the space in half. **AUROC**

The advantage of using ROC curves and the AUROC is that these measurements evaluate the classifier overall, rather than at a particular posterior probability threshold that may be misleading.

[9] Statistical classifiers do not tend to make certain predictions, i.e., with $p = 1$.

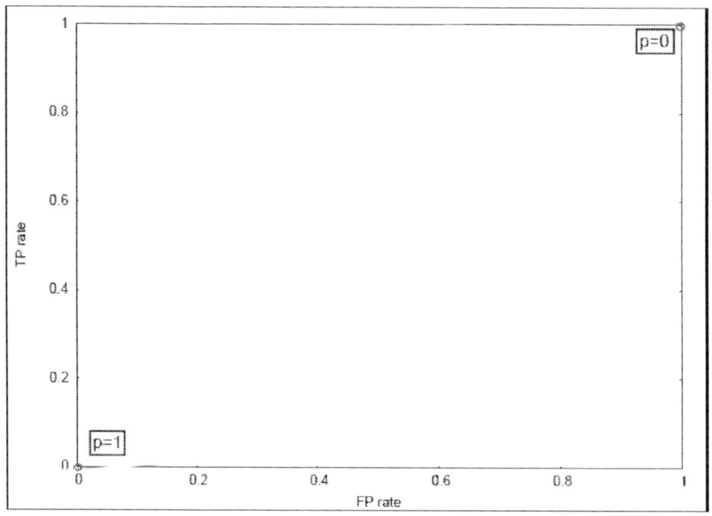

Figure 2.1: The Cartesian quadrant in which an ROC curve is plotted. The first point with threshold p=1 is the origin. The last point with threshold p=0 is the upper right corner.

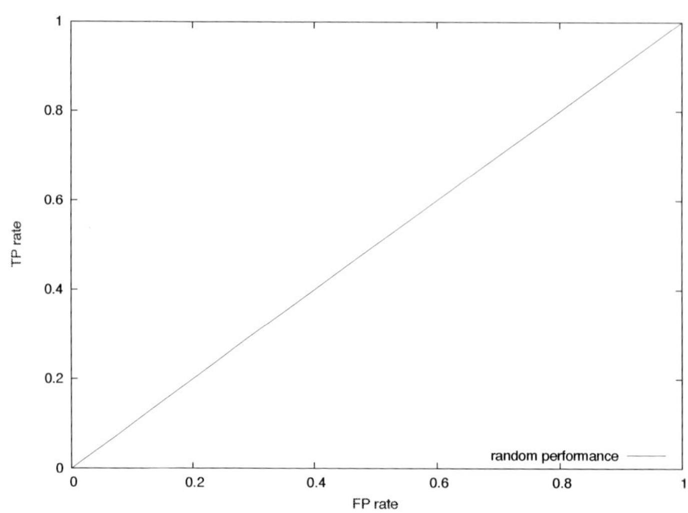

Figure 2.2: ROC curve indicating random performance.

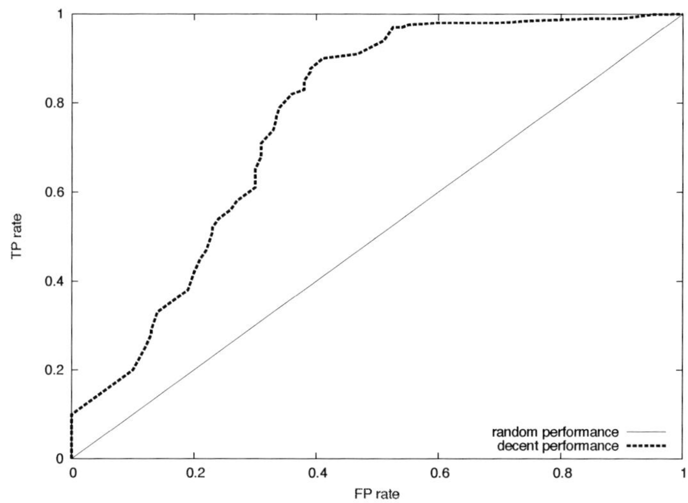

Figure 2.3: ROC curve indicating better performance.

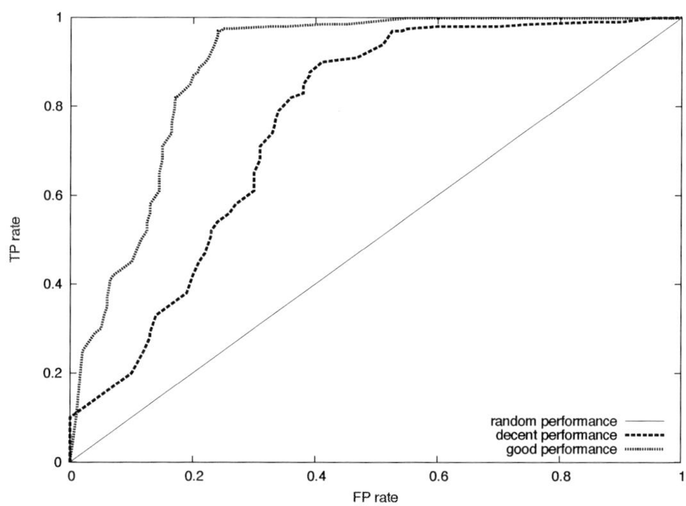

Figure 2.4: ROC curve indicating good performance.

2.3 EVALUATION METRICS FOR SUMMARIZATION

intrinsic vs. extrinsic

Summarization evaluation techniques can generally be classified as *intrinsic* or *extrinsic* [Jones and Galliers, 1995]. Intrinsic metrics evaluate the actual information content of a summary, usually by comparing it either with gold-standard human summaries or with the full document source. Extrinsic metrics, on the other hand, evaluate the usefulness of the summary in performing a real-world task, e.g., using the summary to categorize and file the entire document. For that reason, these are also sometimes called *task-based evaluations*. Most summarization work to date has relied much more heavily on intrinsic measures than extrinsic measures, for the primary reason that such evaluations are more easily replicable and subsequently more useful for development purposes.

2.3.1 INTRINSIC SUMMARIZATION EVALUATION

A definitive overview of summarization evaluation techniques is difficult if not impossible, as the summarization community has never fully agreed on an intrinsic evaluation framework and researchers have tended to rely on their own in-house metrics. The annual Text Analysis Conference[10] has, however, helped to standardize intrinsic evaluations. The submitted outputs of various summarization systems are rated by human judges according to *responsiveness*, or how well a summary meets the information need of a provided query, and various linguistic criteria such as readability.

ROUGE

Carrying out such large-scale human evaluations is feasible for a popular annual workshop, such as TAC, but less so for a researcher aiming to rapidly develop and evaluate a system. In recent years, a suite of automatic evaluation metrics under the name ROUGE has become increasingly popular [Lin and Hovy, 2003]. ROUGE in turn is a variation of BLEU [Papineni et al., 2001], a machine translation evaluation tool. BLEU is based on comparing n-gram overlap between machine translations and multiple gold-standard human translations and is precision-based (with a brevity penalty). An n-gram is simply a sequence of *n* words, so that all the 1-grams (or unigrams) of a document are all the single words of the document, all the 2-grams (or bigrams) of a document are all the sequences of two words, and so on.

n-grams

ROUGE was developed essentially as a recall-based version of BLEU (with a verbosity penalty), in which a system-generated summary is compared with multiple gold-standard human summaries, and the more that the n-grams from the gold-standard summaries also appear in the system summary, the higher the recall score is. The most recent versions of ROUGE also calculate precision and F-score. Basically, the more that the n-grams of the system summary also appear in the gold-standard summaries, the higher the precision is.

skip bigrams

There are several metrics within the ROUGE suite, but the most widely used are ROUGE-2 and ROUGE-SU4, the former of which calculates bigram overlap and the latter of which calculates skip bigram overlap with up to four intervening terms. The following pair of sentences illustrates bigram overlap:

- We can't afford that battery if we want to meet our *final budget*.

[10] `http://www.nist.gov/tac/workshop/`

- The group presented their *final budget*.

The first sentence is from a system summary and the second sentence is from a gold-standard human summary of the document. We can see that the bigram *final budget* occurs in each sentence, so we say that there is a bi-gram overlap between this sentence and the gold-standard. If we permit intervening terms between the words of the bigram, we can identify further overlaps, which are called skip bigram overlaps. The following pair of sentences illustrates skip bigram overlap:

- So let's look at the *final* revised *budget*.
- The group presented their *final budget*.

Here, there is a skip bigram overlap, where one intervening term occurs between *final* and *budget*.

Lin [2004] provided evidence that the ROUGE-2 and ROUGE-SU4 metrics correlate well with human evaluations for several years' worth of DUC data. Subsequent research has yielded mixed results concerning ROUGE correlations with human evaluations [Dorr et al., 2004, 2005, Liu and Liu, 2010, Murray et al., 2005b, 2006], but ROUGE has become an official metric of the Text Analysis Conference and is increasingly relied upon by researchers, allowing them to directly compare summarization results on given datasets.

Basic Elements

The creators of ROUGE have also developed the Basic Elements evaluation suite [Hovy et al., 2006], which attempts to remedy the drawbacks of relying on n-gram units or sentence units for comparing machine summaries to reference summaries. Instead of relying on n-grams like ROUGE does, this evaluation framework uses units called Basic Elements, which are defined in the most simple case as either heads of major syntactic constituents (a single item) or relations between BE-heads and their modifiers (a triple of head, modifier, and relation). The head of a syntactic constituent, e.g., a Noun Phrase (NP) is the word which determines the syntactic category of the constituent and typically conveys the core meaning of the constituent. In contrast, the modifiers of the head simply modify that meaning by specifying values for relevant semantic relations. For instance, *book* is the head of the NP *An interesting book*, while *interesting* is a modifier of the head, which specifies the "quality" of the book. Here, we show an example sentence and some of the associated BE triples < head, modifier, relation >, with semantic relations (in uppercase) determined by a semantic role labeler [Hovy et al., 2005]:

Two Libyans were indicted for the Lockerbie bombing in 1991.

For the NP *Two Libyans*, with head *Libyans*, we have the BE:

- <Libyans|two|CARDINAL>

For the whole sentence, with head *indicted*, we have three BEs:

- <indicted|Libyans|ACCUSED>

- <indicted|bombing|CRIME>
- <indicted|1991|TIME>

The advantage of Basic Elements is that it features a deeper semantic analysis than simple n-gram evaluation so that matches need not be superficial, but the disadvantage is that it relies on parsing and pruning, which can be error-prone for noisy data such as speech and blogs. Like ROUGE, Basic Elements is not a single evaluation metric. Rather it consists of numerous modules relating to three evaluation steps of *breaking*, *matching* and *scoring*, which correlate to locating the basic elements, matching similar basic elements, and scoring the summaries, respectively. Basic Elements can be seen as a generalization of ROUGE, with ROUGE being the special case where the basic elements are n-grams.

Pyramid method

Semantic Content Units

The Pyramid method [Nenkova and Passonneau, 2004] uses variable-length sub-sentential units for comparing machine summaries to human model summaries. These *Semantic Content Units* (SCUs) are derived by having human annotators analyze multiple human model summaries for units of meaning. Each SCU is roughly equivalent to a concept, though SCU itself is not formally defined. Each SCU can have many different surface realizations. For example, the following two sentences relate to the same SCU:

- *They decided to use bluetooth.*
- *The final design included bluetooth.*

The label for this SCU might be *The remote control used blue-tooth*. Each SCU is associated with a weight relating to how many model summaries it occurs in. For instance, Figure 2.5 shows an example in which we have five model summaries and each model summary contains a subset of six SCUs. In this example, the weight for SCU_1 will be 4 (because it appears in four model summaries), the weight for SCU_2 will also be 4, while the weight for SCU_3 will be only 3, etc. These varying weights lend the model the pyramid structure, with a small number of SCUs occurring in many model summaries and most SCUs appearing in only a few model summaries. Machine summaries are then annotated for SCUs as well and can be scored based on the sum of SCU weights compared with the sum of SCU weights for an optimal summary. Figure 2.6 shows a Pyramid for our example in Figure 2.5, containing two SCUs of weight 4 (SCU_1, SCU_2), and four SCUs of weight 3 (SCU_3, .., SCU_6), and two possible optimal summaries containing four SCUs are indicated. These summaries are optimal because they each contain all of the SCUs of weight 4, the highest weight level, and the remaining SCUs from weight 3, the next highest level of the Pyramid.

Using the SCU annotation, one can calculate both precision-based and recall-based summary scores for a given machine summary. For instance, a machine summary containing the four SCUs (SCU_1, SCU_3, SCU_4, SCU_6) would have a precision of 13/14, i.e., the sum of the weights of the SCUs contained in the summary ($4 + 3 + 3 + 3$), divided by the sum of the weight of an optimal summary containing the same number of SCUs ($4 + 4 + 3 + 3$). In contrast, in the recall-based Pyramid score, instead of comparing the machine summary with the ideal summary containing the

Five Summary Models

	A	B	C	D	E	SCU weight
SCU_1	X	X	X	X		4
SCU_2	X	X	X	X		4
SCU_3	X	X			X	3
SCU_4		X	X		X	3
SCU_5	X	X	X			3
SCU_6	X			X	X	3
Num. of SCUs In model summary	5	5	4	3	3	

Figure 2.5: Sample SCU annotation for five model summaries. The matrix shows what SCUs are expressed by each model summary. A cross in a cell indicates that the model summary in the corresponding column is expressing the SCU in the corresponding row. The column on the right show the weight of each SCU, while the row at the bottom show how many SCUs are expressed in each model summary.

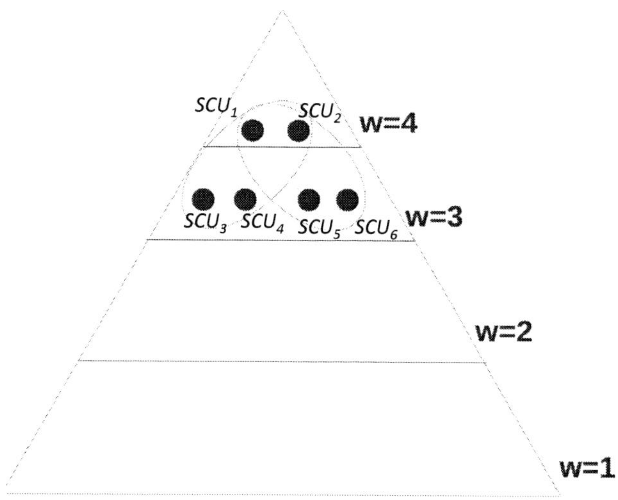

Figure 2.6: Pyramid representation for our example in Figure 2.5. The two ovals encircle two possible optimal summaries with four SCUs. This figure was inspired by a similar figure in Nenkova and Passonneau [2004].

same number of SCUs, we would compare it with an ideal summary containing the average number of SCUs in all the human model summaries used to create the Pyramid model. For instance, in our running example, a machine summary containing the three SCUs (SCU_1, SCU_3, SCU_4) would have a precision of 10/11, but a recall of 10/14, because the average number of SCUs in the model summaries is 4 (see Figure 2.5), and as we have already seen the sum of the weights of an ideal summary of length four is $(4 + 4 + 3 + 3) = 14$.

The advantage of the Pyramid method is that it uses content units of variable length and assigns weights to them by importance according to occurrence in model summaries, but the disadvantage is that the scheme requires a great deal of human annotation since every new machine summary must be annotated for SCUs. Pyramid was used as part of the DUC 2005 evaluation, with numerous institutions taking part in the peer annotation step, and while the submitted peer annotations required a substantial amount of corrections, Nenkova et al. [2007] reported acceptable levels for inter-annotator agreement.

Galley [2006] introduced a matching constraint for the Pyramid method when applied to meeting transcripts; namely, that when comparing machine extracts to model extracts, SCUs are only considered to match if they originate from the same sentence in the transcript. This was done to account for the fact that sentences might be superficially similar in each having a particular SCU, but nevertheless have much different overall meanings.

weighted F-score

The weighted F-score metric [Murray et al., 2006] is analogous to the Pyramid method, but with full sentences as the SCUs. This evaluation metric relies on human gold-standard abstracts, multiple human extracts, and the many-to-many mapping between the abstracts and extracts as described in Section 2.1.1. The idea is that document sentences are weighted according to how often they are linked to an abstract sentence, analogous to weighted Pyramid SCUs. The metric was originally precision-based but was later extended to weighted precision/recall/F-score. The advantage of the scheme is that once the model annotations have been completed, new machine summaries can be easily and quickly evaluated, but the disadvantage is that it is limited to evaluating extractive summaries and works only at the dialogue act level.

The challenge with evaluating summaries intrinsically is that there is not normally a single best summary for a given source document, as illustrated by the low κ scores between human annotators. Given the same input, human judges will often exhibit low agreement in the units they select [Mani, 2001b, Mani et al., 1999]. In early work on automatic text summarization, Rath et al. [1961] showed that even a single judge who summarizes a document once and then summarizes it again several weeks later will often create two very different summaries (in that specific case, judges could only remember which sentences they had previously selected 42.5% of the time). With many annotation tasks, such as dialogue act labeling for example, one can expect high inter-annotator agreement, but summarization annotation is clearly a more difficult task. As Mani et al. [1999] pointed out, there are similar problems regarding the evaluation of other NLP technologies that may have more than one acceptable output, such as natural language generation and machine translation. The metrics described above have various ways of addressing this challenge, relying generally on multiple references. With ROUGE, n-gram overlap between a machine summary and multiple human references is calculated, and it is assumed that a good machine summary will contain certain elements of each reference. With Pyramid, the SCUs are weighted based on how many summaries they occur in, and with weighted F-score, we rely on multiple annotators' links between extracts and abstracts. Teufel and van Halteren [2004] and Nenkova et al. [2007] discussed the issue of how many references are needed to create reliable scores, but the crucial point is that there is no such thing as a single best summary and multiple gold-standard reference summaries are desirable. As Galley [2006] observed, the challenge is not low inter-annotator agreement itself but in using evaluation metrics that account for the diversity in reference summaries.

This has been a necessarily incomplete overview of summarization metrics, as many in-house metrics have proliferated over the years and during that time there was not widespread agreement on which metrics to use. This was a research bottleneck, as it meant that researchers could not easily compare their results with one another. This is less of a problem now, as the community has largely adopted ROUGE and Pyramid as standard metrics. We have also focused on generally applicable metrics in this section, and so have ignored metrics such *summary accuracy* [Zechner and Waibel, 2000] which are speech-specific by incorporating speech recognition error rate. Of all the metrics we have described here, each has advantages and disadvantages. What metrics like ROUGE and weighted precision have in common is that there is an initial stage of manually creating model summaries, and subsequently new machine summaries can be quickly and automatically evaluated. In contrast, Pyramid evaluation requires additional manual annotation of machine summaries. On the other hand, an evaluation scheme like Pyramid operates at a more meaningful level of granularity compared to using n-grams or entire sentences since an SCU roughly represents a concept that can be realized in many surface forms. What all these schemes have in common is replicability, being able to reproduce the results once the relevant annotations have been done, which is not feasible when simply enlisting human judges to conduct subjective evaluations of summary informativeness or quality. Such human evaluations are very useful for periodic large-scale evaluation of summarization systems, however, and crucial for ensuring that automatic or semi-automatic metrics correlate with human judgements or real-world utility.

2.3.2 EXTRINSIC SUMMARIZATION EVALUATION

While intrinsic evaluation metrics are essential for expediting development and can be easily replicated, they should be chosen according to whether they are good predictors for extrinsic usefulness, e.g., whether they correlate to a measure of real-world usefulness. Evaluating in comparison to human gold-standard annotations is sensible and practical, but ultimately all summarization work is done for the purpose of facilitating some task and should be evaluated in the context of that task. As Sparck-Jones has said, "it is impossible to evaluate summaries properly without knowing what they are for" [Jones, 1999]. Ideally, even evaluation measures that compare a system-generated summary with a full source document or a model summary would do so with regards to use constraints.

One popular extrinsic evaluation has been the *relevance assessment* task [Mani, 2001b] . With **relevance assessment** relevance assessment, a person is presented with a description of a topic or event and then must decide whether a given document (which could be a summary or a full-text) is relevant to that topic or event. Such evaluations have been used for a number of years and on a variety of projects [Dang, 2005, Jing et al., 1998, Mani et al., 1999]. Due to issues of low inter-annotator agreement on such tasks, Dorr et al. [2005] proposed a new evaluation scheme that compares the relevance judgement of an annotator given a full text with that same annotator given a condensed text.

reading comprehension

A second type of extrinsic evaluation for summarization is the *reading comprehension* task [Hirschman et al., 1999, Mani, 2001b, Morris et al., 1992]. With a reading comprehension task, a user is given either a full source or a summary text and is then given a multiple-choice test relating to information from the full source. One can then compare how well users perform in term of the quality of their answers and the amount of time to produce them, when given only the summary compared with the full source document. This evaluation framework relies on the assumption that truly informative summaries should be able to act as substitutes for the full source document. This does not hold true for certain classes of summaries such as query-based or indicative summaries (as defined in Chapter 1), which are not intended to convey all of the important information of the source document.

decision audit

A decision audit task [Murray et al., 2009] has been proposed for meeting summarization, and we argue that it could be applied to email summarization as well. In this task, a user must determine which way a group decided on a particular issue and furthermore what the decision-making process was. They are presented with the transcripts of the group's meetings as well as summaries of each meeting and must find the relevant information pertaining to that decision in a limited timeframe. They then write a synopsis of the decision-making process. This synopsis is then evaluated by human judges as to its correctness. By instrumenting the meeting browser, one can also inspect where the user clicked, how frequently they used the summary, whether they played the audio, and so on. Murray et al. [2009] carried out a decision audit evaluation to compare extractive and abstractive summaries and to assess the impact of ASR errors.

Not all extrinsic summarization evaluations involve using the summaries to aid a person performing a task; the summaries could also be used to aid a *system* automatically performing a task. For example, one might be able to improve the precision or recall of a document classification system by first generating summaries of the documents in the collection. We could then evaluate the summarizer by running the document classifier with and without the summarization component [Mihalcea and Hassan, 2005].

2.3.3 LINGUISTIC QUALITY EVALUATION

linguistic quality

A final type of evaluation we will discuss are evaluations of *readability* or *linguistic quality*. This entails scoring the summaries according to fluency, coherence, grammaticality or general readability. It is possible for a summary to be very relevant and informative but to be nearly unreadable to a user, and intrinsic measures such as ROUGE and Pyramid cannot capture that distinction. Typically, we must enlist actual users to make such readability judgments. Linguistic quality assessments are standard evaluations for the TAC/DUC summarization challenges, with five categories corresponding to *grammaticality*, *non-redundancy*, *referential clarity*, *focus* and *structure and coherence*[11].

The reasons why a summary might be informative but still score poorly on readability are diverse; for extractive summaries, it may be the case that the source documents feature very noisy or ungrammatical text. This is particularly an issue with conversational data, where sentences may contain filled pauses, false starts, misspellings and sentence fragments. For example, Sentence 1 below features a false start, but this has been repaired by the summarization system in Sentence 2, leading to better readability:

- *So you will have - Baba and David Jordan, you will have to work together on the prototype.*
- *Baba and David Jordan, you will have to work together on the prototype.*

For abstractive summaries, readability and linguistic quality will largely depend on the quality of the language generation component. If the abstracts are lacking in lexical diversity or do not properly handle anaphora (expressions referring to other expressions, e.g., pronouns), to give just two examples, they will likely be scored poorly on linguistic quality. **anaphora**

2.3.4 EVALUATION METRICS FOR SUMMARIZATION: A FINAL OVERVIEW

Figure 2.7 places the major evaluation methods we have discussed onto two axes, one describing how automated the evaluation is and one describing how deeply the summaries are analyzed. By automated vs. manual, we are indicating how much manual intervention needs to be done in order to evaluate a newly generated summary. Evaluation methods such as ROUGE and precision/recall/F-score require only an initial manual generation of gold-standard extracts or abstracts, and subsequently new summaries are evaluated in a fully automatic fashion. In contrast, the Pyramid evaluation requires annotation of each new summary, and extrinsic evaluations that measure how well the summaries aid a user in performing a task require a great deal of manual intervention, such as recruiting participants, designing the study, analyzing the results, etc. An extrinsic evaluation setup that measures how automatic summarization improves an information retrieval task, on the other hand, might be automated and easily replicable.

By shallow vs. deep, we are indicating whether the evaluation methods are analyzing the summaries at a superficial, surface level or at a deeper level corresponding to meaning or utility. ROUGE and sentence precision/recall/F-score are both fairly shallow, measuring n-gram overlap and sentence overlap with gold-standard summaries, respectively. Pyramid and Basic Elements both operate at a more semantic, conceptual level, while extrinsic summaries go beyond meaning to measure actual utility.

[11] `http://www-nlpir.nist.gov/projects/duc/duc2007/quality-questions.txt`

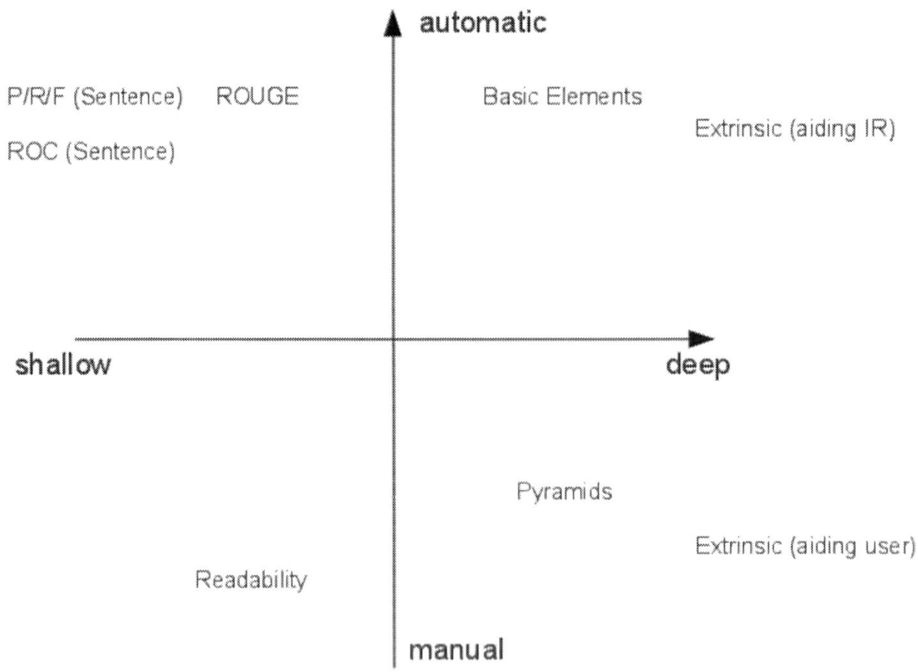

Figure 2.7: Types of summarization evaluation.

2.4 CONCLUSION

In this chapter, we reviewed several, although certainly not all, of the conversation corpora used by researchers. We also described and compared the annotation schemes applied to such corpora. Likewise, we presented some of the most popular evaluation metrics for summarization and text mining of conversations, but it is not uncommon for researchers to devise their own in-house evaluation metrics—a trend that can make it difficult to exactly compare contrasting approaches in the literature. When presenting case studies in the following chapters, we will often report evaluation metrics based simply on what was published in the original work.

2.5 IMPORTANT POINTS

- Large collections of documents, often annotated, are called *corpora*.

- Datasets that have been *annotated* or *coded* have been manually labeled for phenomena of interest.
- Two evaluation metrics widely used for mining tasks are *precision/recall/F-score* and *ROC curves*.
- ROUGE, Basic Elements and Pyramid are examples of *intrinsic* summarization evaluation metrics that measure the information content of a summary.
- *Extrinsic* summarization evaluation metrics measure how useful a summary is for a particular task.

2.6 FURTHER READING

The NLTK book, available online[12], has a chapter on "managing linguistic data," providing information on how to create, format and document linguistic resources such as an annotated corpus. NLTK itself[13] contains many corpora, some of which are annotated.

Mani [2001b] provides a very good overview of summarization evaluation issues. While that paper predates current evaluation toolkits such as ROUGE and Pyramid, it features a high-level discussion of evaluation issues that is still relevant today, e.g., in its discussion of intrinsic vs. extrinsic approaches.

[12] `http://www.nltk.org/book`
[13] `http://www.nltk.org/`

CHAPTER 3

Mining Text Conversations

3.1 INTRODUCTION

In this chapter, we describe several text mining techniques that can be applied to conversations and explain why they might be useful for other tasks such as summarization. We can see from the example conversation in Chapter 1 that there are numerous questions a person might ask if they wanted quickly to understand the discussion: *What was the topic of discussion?*, *What was proposed?*, *What opinions were expressed?*, *What was finally decided?* The mining techniques described in this chapter attempt to answer these and other questions.

We first discuss topic modeling, comprising the two related tasks of topic segmentation and topic labeling. We then describe the broad field of sentiment and subjectivity detection and several specific sentiment tasks. After that, we cover tasks related to mining the conversation structure, including dialogue act classification, decision and action item detection, and extraction of thread structure. In each section, we give examples of current work on conversational data. We conclude the chapter by giving pointers to further reading on each area of interest.

3.2 TOPIC MODELING: TOPIC SEGMENTATION AND TOPIC LABELING

Any document spanning more than a few sentences is very likely to cover more than one topic. For instance, Hearst [1997] reports that in her corpus of expository text the end of each paragraph has approximately a 40% chance of being a topic boundary.

The task of topic modeling aims to capture the topical structure of a document (or a collection of documents) by identifying what topics are discussed in the text, and which portions of text correspond to which topics. When the goal is limited to splitting the input document(s) into segments, where each segment is about a single topic, we talk about *topic segmentation*. In contrast, complete topic modeling includes both topic segmentation and *topic labeling*, in which all the topics covered in the input document(s) are labeled with informative descriptions, ranging from simple sets of words to more informative phrases.

topic segmentation

topic labeling

As an example, Table 3.1 shows a possible multi-paragraph topic model for a 23-paragraph article about the exploration of Venus by the Magellan space probe[1].

hierarchical topics

Topic models can be flat or hierarchical. In a flat topic model, text is modeled as a sequence of topical segments with no further decomposition, while in a hierarchical topic model segments can be

[1] Source: http://people.ischool.berkeley.edu/~hearst/research/tiling.html

Table 3.1: Sample human generated topic model of an article on the exploration of Venus by the Magellan space probe. The reader split the document into ten segments. For each segment, the numeric range indicates the article paragraphs comprising that segment, while the label specifies the reader description for the segment.

Article Paragraphs	Reader Description for the Segment
1-2	*Intro to Magellan space probe*
3-4	*Intro to Venus*
5-7	*Lack of craters*
8-11	*Evidence of volcanic action*
12-15	*River Styx*
16-18	*Crustal spreading*
19-21	*Recent volcanism*
22-23	*Future of Magellan*

further divided into subtopics. For instance, in a hierarchical model of the example in Table 3.1, the high level topic "Recent volcanism" could be further divided into "Geographical distribution of active volcanoes" and "Evidence from images". Arguably, a hierarchical topic model would more effectively support document browsing, retrieval and summarization, because of the richer and finer-grained information it provides.

The difficulty of topic modeling varies in different text domains. In edited monologues, such as books and articles, topic modeling is considered relatively simple, as material on different topics typically comes already organized in different chapters, sections and paragraphs, which reflect the topical structure.

In contrast, in less formal documents, including text conversations, where material is unedited and less explicitly organized, topic modeling becomes much more complex. For instance, in a conversation the beginning of a topic and the end of the previous one often overlap and a topic may be introduced multiple times before it becomes the focus of the discussion.

To address these challenges, work on topic modeling of conversations often consists of adapting and extending methods devised for generic text. In this section, we will first review these general methods and then show how they have been adapted to deal with conversations.

3.2.1 TOPIC MODELING OF GENERIC TEXT

Cohesion Based Segmentation Approaches: A key intuition behind many approaches to topic segmentation is that sentences within a segment are more connected to each other than to sentences in other segments. The strength of connection between two sentences is called cohesion in linguistics and it is determined by how close the words in the two sentences are (i.e., lexical cohesion) and by the use of other linguistic devices, including primarily pronouns.

lexical cohesion

For instance, if you look at these three sentences:

1. "Ciro is the best pizza maker in town."
2. "He serves super fresh ingredients on a very thin crust pizza!"
3. "They do not think Incendies is still playing at a movie theater on Granville."

Sentence 1 and sentence 2 are very cohesive. The word "pizza" mentioned in 1 is repeated in 2 and the subject of 1 "Ciro" agrees in number and gender with the pronoun "He" in 2. Furthermore, the word "pizza" in 1 is also semantically close to the words "crust" and "ingredients" in 2, because the crust is *a part of* a pizza, and because a pizza, being a type of food, is *made of* ingredients. In contrast, sentence 1 and sentence 3 are not cohesive, since their words are semantically quite distant and the subject of sentence 1 "Ciro" does not agree in number with the pronoun "They" in sentence 3.

One of the first and most influential methods for topic segmentation based on lexical cohesion is **TextTiling**, which was developed in the 90s [Hearst, 1997].

TextTiling

An extremely simplified version of the TextTiling algorithm can be described as follows (see Jurafsky and Martin [2008] Chapter 21 for details). Two adjacent sliding windows covering blocks of (let us say) five sentences are moved down on the target document. At the onset, the first window covers the first five sentences of the document, while the second will cover the block from sentence 6 to sentence 10 (see top of Figure 3.1). At each iteration, the two windows are slid one sentence down. So, at the second step the two windows will cover the 2-6 and 7-11 sentence blocks, respectively. At each iteration, a lexical cohesion score between the two sentence blocks is also computed and assigned to the gap between the two blocks. Such a score intuitively measures to what extent the words in the two blocks overlap (see Chapter 4 for a discussion of cosine similarity scores). Once the end of the document is reached, the algorithm looks at the plot of the cohesion scores collected at each gap between two adjacent blocks. Whenever there is a deep valley in the cohesion function, the gap corresponding to the bottom of the valley (where the blocks were minimally similar) is returned as a plausible topic segment boundary. For instance, if the plot in Figure 3.1 represented the cohesion scores computed by TextTiling on a given document, gaps 11 and 25 would be good candidates as segment boundaries. Notice that selecting the bottom of the deep valleys of the cohesion function matches the assumption that text from two different segments should be minimally cohesive.

The basic ideas behind TextTiling have been later refined in more sophisticated algorithms (e.g., see Choi [2000] and Utiyama and Isahara [2001]), which still represent challenging baselines for more recent approaches.

Probabilistic Topic Modeling: A novel approach to topic modeling called Latent Dirichlet Allocation (LDA) was presented in Blei et al. [2003] (see Blei and Lafferty [2009] for a gentle

Latent Dirichlet Allocation (LDA)

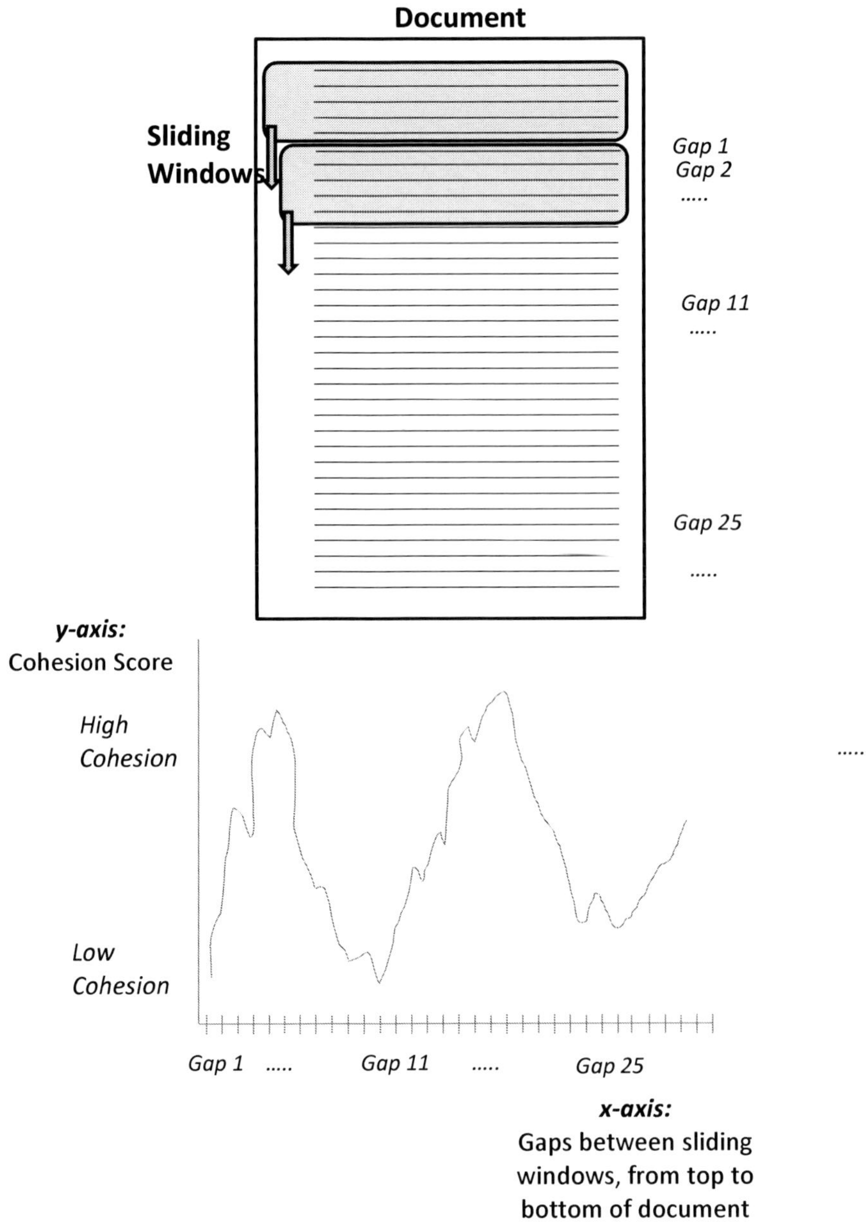

Figure 3.1: TextTiling topic modeling algorithm. A cohesion score is measured at each sentence gap as two sliding windows are moved down on a target document. Bottoms of deep valleys in the cohesion score function indicate segment boundaries.

introduction). The key intuition behind LDA is to model the generation of a document (or of a collection of documents) as a stochastic process in which words are selected by sampling some discrete probability distributions. More specifically, we assume that our documents are about a set of topics, each document is a multinomial distribution over topics, and each topic is a multinomial distribution over words. Figure 3.2 shows examples of these distributions for a toy LDA model involving two documents, three topics and a vocabulary of one thousand words.

Sample Multinomial Distributions over Topics

		Topic T_1	Topic T_2	Topic T_3
θ^1	Document 1	.1	.7	.2
θ^2	Document 2	.3	.3	.4

Sample Multinomial Distributions over Words

		Word w_1	Word w_2	......	Word w_{1000}
ϕ^1	Topic T_1	.001	.021	......	.00006
ϕ^2	Topic T_2	.0021	.006	......	.03
ϕ^3	Topic T_3	.0065	.0043	......	.009

Sample Dictionary

Word w_1	Word w_2	......	Word w_{1000}
ability	*abrasion*	......	*youth*

Figure 3.2: Probability distributions for a sample LDA Topic Model of two documents, involving three topics and a dictionary of one thousand words.

Assuming that the variable z ranges over the topics (T_1, T_2, T_3 in our example), and the variable w ranges over the words (w_1, ..., w_{1000} in the example), we can refer to the topic-word distribution and document-topic distribution as $\phi^{(j)} = P(w|z_i = j)$ (one for each topic) and $\theta^{(d)} = P(z)$ (one for each document). Now we can more formally specify the stochastic process that generates all the words of document d in a collection: it consists of repeated samples from $\theta^{(d)} = P(z)$ to get a topic and from $\phi^{(j)} = P(w|z_i = j)$ to get a word given that topic.

For the more statistically inclined, an LDA model specifies the following distribution over words within a document, by combining ϕ and θ distributions:

$$P(w_i) = \sum_{j=1}^{T} P(w_i|z_i = j)P(z_i = j) \text{ ,}$$

where T is the number of topics. $P(w_i|z_i = j)$ is the probability of word w_i under topic j and $P(z_i = j)$ is the probability that j^{th} topic was sampled for the ith word token.

The power of an LDA formalization is that such a model (i.e., all the probability distributions) can be effectively learned for any given set of documents. Not only is there an efficient method to estimate ϕ and θ (variational EM), but one can also directly estimate (by Gibbs sampling) the posterior distribution over $z = P(z_i = j|w_i)$; namely, a topic assignment for words which specifies how likely is a topic given a certain word.

By assuming the words in a sentence occur independently, a topic assignment for words allows us to also compute a topic assignment for sentences as follows: the topic for a given sentence s should be the one with the highest probability given all the words in s, formally:

$$j^* = argmax_j P(z_i = j|s),$$

where

$$P(z_i = j|s) = P(z_i = j|w_1, ..., w_n) = \prod_{w_i \in s} P(z_i = j|w_i) .$$

As a final step, LDA enables topic segmentation. Once we have assigned to each sentence its most likely topic, blocks of adjacent sentences sharing the same topic will constitute the topical segments.

If the reader is familiar with graphical models (e.g., Bayesian Networks), an LDA model is a probabilistic generative graphical model that formally describes the generation of all the documents in a collection. Figure 3.3 shows the model in plate notation[2] for D documents, T topics and N_d words in each document d. The only variables that are observed in the graphical model (grayed in the Figure) are the ones corresponding to the words in the documents, while all the other variables are hidden and include z, a topic assignment for each word in each document, the distributions $\theta^{(d)}$ and $\phi^{(j)}$, as well as Dirichlet priors for those, α and η, respectively, (from which LDA gets its name).

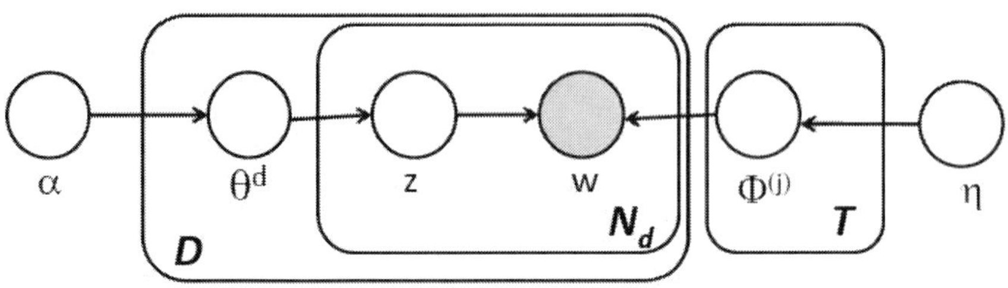

Figure 3.3: Graphical model for LDA in plate notation. The only observed variables are the w nodes corresponding to words. As the plate notation shows, the model contains a node for each word, in each document of the collection.

[2]Plate notation is a concise way to show complex graphical models, in which some subgraphs are replicated for each element in a set. More specifically, each subgraph encircled in a box must be replicated for each element of the set denoted by the box label. For example, in Figure 3.3 the subgraph contained in the biggest box must be replicated for each document in the set D. For more details on the plate notation, see Poole and Mackworth [2010].

One often mentioned limitation of LDA is its inability to choose the optimal number of topics. Possible approaches to this problem are discussed in Blei and Lafferty [2009].

Supervised Classification: TextTiling and LDA are examples of unsupervised techniques, since they do not need to be trained on a corpus annotated with segmentation and topics. However, if such a corpus is available for a specific domain (e.g., news article, email), supervised machine learning approaches can be effectively applied to the topic modeling task.

For instance, text segmentation can be framed as a binary classification task in which given any two adjacent blocks of sentences the classifier would predict whether the gap between them is a segment boundary or not. Several features of the two sentence blocks have been considered in the literature, including word overlap and cosine word similarity (see Chapter 4) between the blocks, whether the terms in the two blocks refer to the same entities (e.g., like "Ciro" and "He" in sentences 1 and 2 at the beginning of this section), and the presence of discourse markers (also called cue words) at the beginning of the second block. Discourse markers are specific words or phrases such as *"Well"* and *"Let's"* that strongly correlate with the start of a new segment. Finally, as is often the case in machine learning, the output of unsupervised techniques (e.g., the estimates of LDA) can be added to the feature set.

Topic labeling can also be framed as a classification task. If you have a corpus in which each segment is labeled with its corresponding topic, a classifier can be trained to predict the topic of any give segment. All kinds of text features can be used, including lexical and syntactic ones. In this case, the classification task is a multi-class one, with a class for each topic covered in the corpus. For instance, in a corpus on documents about "natural disasters," a multiclass classifier could be built to identify segments about "effects on the population," "effects on the infrastructure," "plan/cost of reconstruction," etc.

Classification (binary or multi-class) is just the simplest way to turn topic modeling into a supervised problem. Since the task essentially involves labeling a sequence of gaps (or sentences), more sophisticated supervised sequence labeling techniques can be applied (e.g., Hidden Markov Models (HMM), Conditional Random Fields (CRF) [Poole and Mackworth, 2010]). **sequence labeling**

All the basic approaches to topic modeling of generic text are summarized in Figure 3.4. As we have seen, topic segmentation can be performed in an unsupervised way by either considering the cohesion between segments (TextTiling), or by learning a probabilistic generative model for the target documents (LDA). LDA, in particular, can be also used for the topic labeling task. For supervised approaches to topic modeling (second column in Figure 3.4), binary and multi-class classification methods, as well as sequence labeling ones can be effectively applied. In the next section, we will discuss how all the approaches summarized in Figure 3.4 have been extended and sometimes combined to perform topic modeling of conversations.

3.2.2 TOPIC MODELING OF CONVERSATIONS

Most previous work on topic modeling of multi-party conversations has focused on meeting transcripts. Only very recently, researchers have started to work with emails and blogs for this task.

We will start our discussion from meetings and then move to other conversational modalities. For each proposal we will point out whether the proposed technique is unsupervised, supervised, or a combination of the two. Unsupervised methods will be characterized with respect to whether they follow a cohesion-based, a probabilistic LDA or other approaches. In contrast, for supervised methods we will give special attention to what features are used by the classifier.

Topic Modeling for Meeting Transcripts The first comprehensive approach to topic segmentation in meetings was presented by Galley et al. [2003]. Their method combines ideas from unsupervised cohesion-based segmentation with a supervised approach. As in TextTiling, their unsupervised **LCSeg** cohesion-based technique, called LCSeg, first computes a cohesion score for each potential segment boundary (every gap between two utterances) and identify as boundaries all the gaps where that function drops significantly and reaches a minimum. One improvement with respect to TextTiling is that their cohesion function is not simply based on shallow word vector similarity, but it also considers the more sophisticated notion of lexical chains (i.e., sequences of related words spanning multiple **lexical chains** sentences [Morris and Hirst, 1991]). Another advantage of LCSeg is that instead of returning a yes/no boundary decision for each potential boundary, it can return a probability estimate, which can later be used effectively in the supervised method.

Their supervised method follows the standard approach we described for generic text, in which topic segmentation is framed as a binary classification task. However, Galley et al., in addition to cohesion-based and discourse marker features, also considered conversational features. Of these, the ones that are not meeting/speech specific and can be applied to other text conversations, include: (i) gaps/pauses/silences between utterances, under the assumption that the longer the gap the more

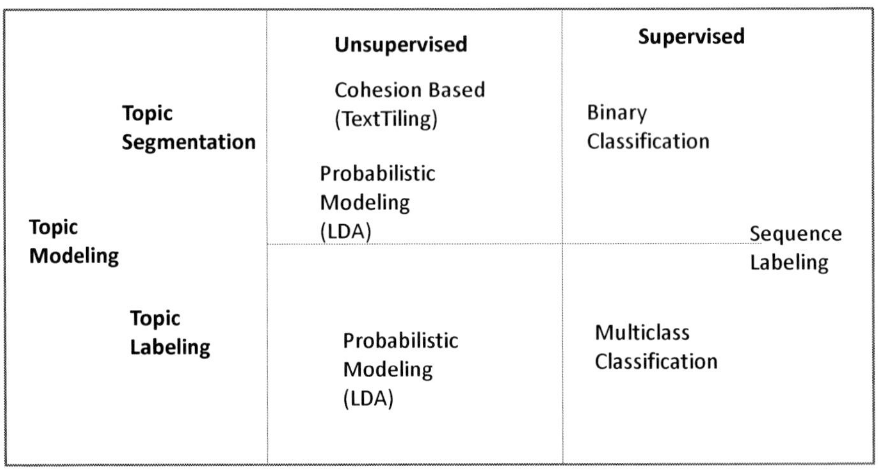

Figure 3.4: Topic modeling approaches for generic text.

likely is that a topic shift occurred; (ii) significant changes in speakership, under the assumption that changes in the amount of activity of the different speakers correlates with changes in topic.

Galley et al. [2003] also performed an evaluation of their supervised approach by training and testing a decision tree classifier[3] on the ICSI Meeting Corpus (see Chapter 2). They found that although cohesion-based features are more critical than conversational ones, the system performs best when all the features are used.

As we mentioned before, the topical structure of a document can be either flat or hierarchical. While in a flat structure the text is simply modeled as a sequence of topical segments with no further decomposition, in a hierarchical topic model segments can be further divided into subtopics. The idea of integrating unsupervised cohesion-based segmentation with a supervised approach has been also applied to detect hierarchical topic models of meetings. By following Galley's et al. approach of combining LCSeg with a set of conversational features, Hsueh et al. [2006] have explored how to perform topic modeling of meetings at different levels of granularity. They start by noticing that a meeting can be often divided into a set of major topics, which can be further divided into more refined sub-topics. For instance, a research project meeting could include as major topics *status-report* and *how to proceed*, and *how to proceed* could be further segmented into *experiment design* and *data collection*.

In their experiments, again on the ICSI meeting corpus transcripts, they compare the performance of different segmentation approaches on the two tasks of identifying macro-topic vs. sub-topic boundaries. Their findings indicate that the two tasks are quite different in this respect. While for predicting major-topic shifts a supervised combination of lexical and conversational features works best, for sub-topic shifts an unsupervised lexical-cohesion based method performs as well as the supervised one.

Topic modeling of conversations can also be framed as a probabilistic modeling problem by extending the basic LDA unsupervised framework. All this work is technically quite sophisticated, so we limit our treatment to the basic ideas and insights. Purver et al. [2006b] present an extension of LDA that explicitly models a topic shift between two utterances with an additional binary hidden variable c_u, one for each utterance, indicating whether there is a shift after that utterance ($c_u = 1$) or not ($c_u = 0$). Using LDA terminology, a topic shift corresponds to a change in the probability distribution over topics.

Figure 3.5 shows the graphical model corresponding to this variation of LDA. In this model, each utterance in the conversation plays the role of a document in a collection. So, what is D in Figure 3.3, here becomes U. By design, the distribution over topics for each utterance is conditioned on c_u. Furthermore, since utterances are sequentially ordered, this model makes the Markov assumption that the distribution over topics of an utterance depends on the distribution over topics of the previous utterance (arrows pointing down connecting the plate for $u - 1$ to the one for u, and the one for u to the one for $u + 1$).

[3] See Poole and Mackworth [2010] for an introduction to decision trees.

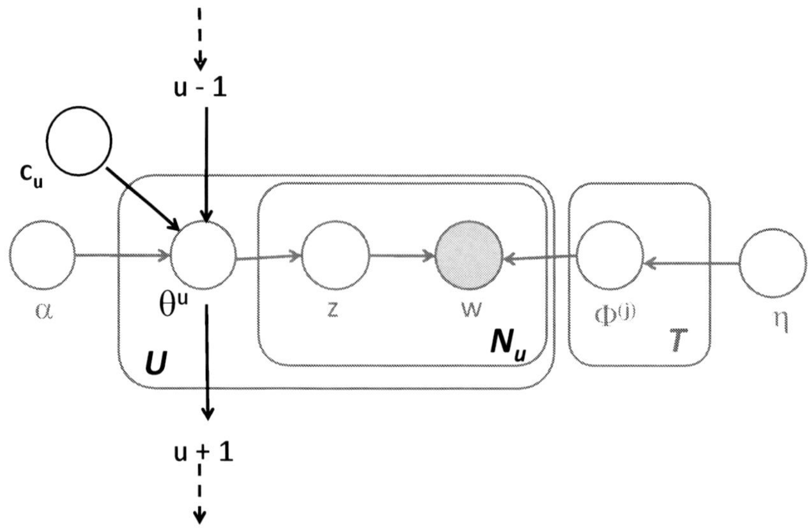

Figure 3.5: Graphical model for a variation of LDA that also performs topic modeling. Additions and changes to the standard LDA model are highlighted in black.

Once the model is learned (in an unsupervised fashion) and it is applied to a conversation, the values predicted for all the c_u variables will provide the topic segmentation for the conversation, with a topic shift every time $c_u = 1$. Since the model is an extensions of LDA, it will also provide labels for the topics. When tested on the ICSI corpus, this extension of LDA performs similarly to LCSeg for topic segmentation, with the additional advantage of providing topic labels that were found intuitively informative by the author and semantically coherent by seven independent human judges.

Other variations of LDA have been recently explored for modeling topics in meeting transcripts. For more information, the interested readers can refer to Huang and Renals [2008] and Georgescul et al. [2008].

Topic Modeling for Microblogs and Email Although most of the approaches to topic modeling of text conversations have been so far developed and tested on meeting transcripts, a few researchers have begun to investigate their application to other conversational modalities. For example, **microblogs** Ramage et al. [2010] study topic modeling for *microblogs* such as Twitter. In particular, they show **Labeled LDA** how a semi-supervised variation of LDA, called labeled LDA, can be effectively applied to Twitter conversations, in spite of the very limited length of Twitter posts - 140 characters or less. Labeled LDA [Ramage et al., 2009] is a generalization of LDA that can be effectively applied in domains in which there is prior knowledge on the topical structure of the documents. In essence, if there are some topics that one cares about, for instance if the documents are already annotated with some topic labels, then it is possible for the LDA to only use those when it learns the topic model.

Figure 3.6 shows the graphical model for Labeled LDA, where Λ represents the set of all possible topics. Since we assume the existence of labels for each document, Λ is observed for each document (grayed in the Figure).

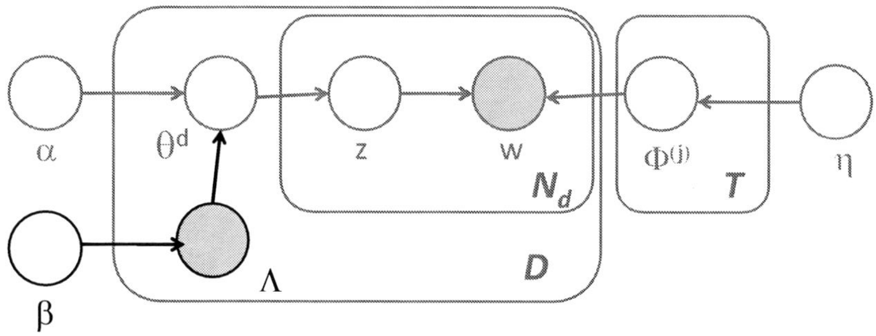

Figure 3.6: Graphical model for labeled LDA. Additions to the standard LDA model are highlighted in black.

To apply Labeled LDA to Twitter, Ramage et al. first conducted a set of structured interviews to identify what are the basic dimensions people consider when they decide what posts to read or what user to follow on Twitter. They found four such dimensions (called 4S): substance topics (about an entity or idea), social topics where language is used towards a social end (e.g., making plans with friends), status topics conveying personal updates, and style topics (e.g., humor or wit). Then, through a rather complex semi-automated process they label a large Twitter dataset with those four labels and run Labeled LDA on it.

The output of this process is a topic model for Twitter conversations that is only based on four topics, namely, the 4S. This model can be applied to any set of tweets. Figure 3.7, from Ramage et al. [2010], shows how the tweets of two sample users can be a visualized in the context of a 4S topic model[4].

Ramage et al. also ran a user study which indicates that the learned topic models would be effective in helping Twitter users to identify the most valuable posts in their current feed, as well as what new users to follow.

Recently, there has also been work on applying topic modeling techniques to *email* conversations, with the limited goal of generating summary keywords for each email message. In an empirical comparison of different unsupervised approaches, LDA has been shown to be the best performer [Dredze et al., 2008]. In essence, once a set of email messages has been modeled with LDA, the best keywords to describe an email are the ones with the highest probability given that email. The probability of each candidate keyword c, given an email e, can be formally computed as:

$P(c|e) = \sum_{j=1}^{T} P(c|z_i)P(z_i|e)$

[4]Additional interactive examples can be explored at `http://twahpic.cloudapp.net/`

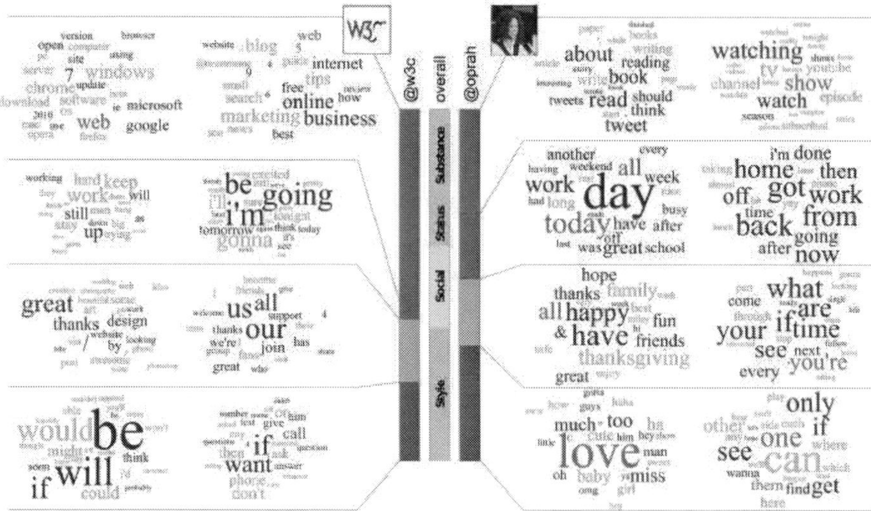

Figure 3.7: Distributions of 4S topics for two users, when compared with the overall average distributions in Twitter. The overall distribution is shown in the middle bar, with Status less frequent than the other three topics, which are similarly distributed. The right bar shows the topic distribution for user @oprah, which is similar to the overall distribution. The bar on the left show the distribution for user @w3c, which is instead substantially different from the other two, as the topic Substances dominates the distribution. Beside the users' topic distributions, common words for each topic are shown as word clouds. Size and shade of the words convey frequency and recency of usage, respectively.

An intuitive way to interpret this formula is that the best keywords for an email e (i.e., the ones with higher $P(c|e)$), are the ones that are highly probable (high $P(c|z_i)$) in the most likely topics for that email message (high $P(z_i|e)$).

Notice that this is a straightforward application of LDA. However, it provides an interesting example of how topic modeling can be used to perform a very basic form of summarization, as for each email, we can compute the summary keywords that best describe the email in the context of a topic model. These short summary descriptions can be used as substitute for the original emails, either to facilitate the user interaction with an email repository, or to improve other email processing tasks. Experiments on the Enron corpus (see Chapter 2) show that these keyword summaries can support more effective automatic email foldering as well as the prediction of an email intended recipients.

Current and Future Trends in Topic Modeling for Text Conversations What are the current open issues in topic modeling for conversations? We expect much more work on asynchronous conversations, including emails and blogs, with particular interest in two questions.

- How can the previous work on meeting transcripts, in terms of corpora and techniques, be leveraged to deal with these different media? For instance, how can domain adaptation techniques [Daumé and Marcu, 2006] be applied?

- How can topic modeling techniques be informed by other text mining tasks that we discuss in this chapter? For instance, how can topic modeling be effectively informed by a finer level analysis of the structure of the conversation? An initial investigation of this idea is presented in Joty et al. [2010].

Current and future work on topic modeling of conversations should leverage publicly available tools. For instance, the Stanford Topic Modeling Toolbox provides a publicly available Java implementation of LDA and Labeled LDA[5].

3.3 SENTIMENT AND SUBJECTIVITY DETECTION

In this section, we first give a general overview of sentiment and subjectivity research on generic text, and then discuss such research as it applies to conversational data.

3.3.1 SENTIMENT DETECTION BACKGROUND

Research related to identifying opinions and subjective language is a wide field with many names: *opinion detection*, *sentiment detection*, *subjectivity detection*, *polarity classification*, and *semantic orientation*, among others. Pang and Lee [2008] give a detailed history of how these names arose and how they have been used since, but argue that they all essentially denote the same field of study. We concur, and more or less use these terms interchangeably. They are all concerned with identifying what people think or feel, in terms of opinions and emotions expressed. Such subjective language contrasts with objective, factual language.

If we again consider the sample conversation from Chapter 1, we can observe that sentences such as *Great idea!*, *I do not like this assignment* and *Too cold this time of the year* all involve the expression of an opinion or emotion held by one of the discussion participants, and contrast with objective, factual sentences such as *By the way I am working on the assignment.* This difference is the basic interest in the field of opinion/sentiment/subjectivity detection.

Using these terms interchangeably is not meant to mask that there are many different tasks within this field of study. One way of differentiating these tasks is to describe them in terms of *granularity*. At the coarsest granularity, we may be asking questions such as *Do people like this political candidate?* and *What do people think of this product?*. At the finest granularity, we may be trying to determine which facets of a product a particular person likes and dislikes, and why. Here, we describe several tasks relating to subjectivity detection, arranged approximately from coarsest to finest granularity:

[5] `http://nlp.stanford.edu/software/tmt/tmt-0.3/`

- **Online Sentiment Classification** Companies, political parties and other institutions are often very interested in what people think of their "product." One way of soliciting feedback and opinions is to carry out a small-scale focus group using specially selected individuals. An attractive alternative is to mine the web, analyzing the opinions of thousands or millions of people who are discussing the products online. These discussions could be reviewer comments on a site such as Amazon.com, or more informal conversations on blogs and in discussion fora. The needs of the institution will determine the level of sentiment granularity required, but often the result will simply be an indication of whether opinions are favourable, unfavorable or mixed– whether the *polarity* is positive or negative—and how strong those opinions are.

polarity

- **Document Sentiment Classification** At times, we want to know whether a single document, such as a product or movie review, is positive or negative. That is, we want to know the polarity of the document. This will also often involve a rating of the polarity strength. Did the reviewer hate the movie, or simply find it boring? The former would be a stronger negative polarity than the latter. This polarity rating will necessarily be a simplification, and sometimes a very crude one. For example, if the reviewer loved the movie script but hated the acting, should we give a "neutral" polarity rating? In any case, deriving a polarity rating will entail losing some information, but note that reviewers themselves often assign a star-rating to the review they authored. Such simplification can be useful and is a key part of sentiment analysis[6].

- **Sentence Sentiment Classification** We may want to dig deeper and identify all the points where somebody was expressing an opinion. This could involve classifying every sentence in a document as subjective or non-subjective. Given the transcript of a meeting where people were discussing the product they were designing, we could run a sentiment classifier over the data and quickly see all the opinions the people involved had about the product, favourable or not. Similarly, we could run a *polarity* classifier over the data that would allow us to extract just the positive or just the negative opinions.

- **Facet Sentiment Classification** Knowing that a person likes or dislikes something is often not enough. We want to determine the *reasons* behind the opinions they hold. Given a product or some other entity, we can identify *facets* of the product and correlate opinions to each individual facet. For example, given the same meeting transcript described above, we could identify facets of the product they are describing such as the size, weight, color and interface of a remote control they are designing, and then describe what people think about each of these specific facets.

facets

There are several dimensions we can use to describe work on subjectivity and sentiment detection. Like most research in text mining and natural language processing, there are *supervised* and *unsupervised* approaches. Most of the tasks listed above lend themselves well to supervised

supervised vs. unsupervised

[6]Indeed, all tasks described in this book involve simplification of some kind.

machine learning techniques, where a classifier can be trained on data labeled as subjective vs. non-subjective or positive vs. negative, at the document or sentence level. Alternatively, semi-supervised or unsupervised algorithms can predict subjectivity and polarity with little or no labeled data; an example of a semi-supervised approach is to use a manually selected set of *seed* words that are known to be subjective or to have a particular polarity, and use those seed words to automatically label sentences or documents. This can lead to the discovery of new subjective words, expansion of the seed set and repetition of the whole process. This would be an example of a *boot-strapping* procedure. **seed words** **boot-strapping**

A related distinction is between *lexicon-based* approaches and *statistical* approaches, though this is less of a theoretical distinction than a reflection of common system implementations. In a lexicon-based approach, there is a dictionary of subjective or polar words, usually associated with numerical scores indicating the strength of the word polarity. For example, the scores may range from -5 to +5, with -5 indicating very negative sentiment (e.g., "terrible") and +5 very positive sentiment (e.g., "awesome"). Given a text, a lexicon-based system identifies words contained in its lexicon and retrieves their word scores. A phrase, sentence or document can be scored, in the simplest case, by summing over its sentiment word scores. **lexicon-based**

SO-Cal [Taboada et al., 2010] (for *Semantic Orientation Calculator*) is a lexicon-based system that is considerably more sophisticated than that and illustrates why simply summing over word scores is not sufficient. To give just one example, such a system must account for *negators* that can weaken or reverse a word's dictionary score. The following three sentences help illustrate this phenomenon: **negators**

1. I love this interface design.
2. I don't love this interface design.
3. I hate this interface design.

It seems clear that Sentence 1 is very positive and Sentence 3 is very negative, as indicated by the words *love* and *hate*, respectively. However, Sentence 2 also contains the word *love*. Based solely on the dictionary scores for the sentiment words, this sentence should therefore be considered positive as well. Of course, we know that the preceding word *don't* negates that positive sentiment, and any system will need to account for this effect. However, if we simply reverse the sentence score due to the presence of the negator, we will end up assigning a very negative score similar to the score for Sentence 3. Intuitively, it seems that Sentence 2 is more ambivalent than Sentence 3 and should have more of a neutral score. For that reasons, systems like SO-Cal make more subtle adjustments to a sentence score when a negator is present.

In contrast, many *statistical systems* do not rely on hand-crafted dictionaries, but rather automatically learn subjective terms or phrases from labeled or unlabeled data. One idea is to build a list of subjective words by identifying the words that occur most frequently in text labeled as being subjective, once stopwords have been removed. Other statistical systems never use an explicit list of subjective or polar words, but rather extract raw lexical features such as unigrams and bigrams and let the machine learning method automatically learn how those features correlate with the positive **statistical systems** and negative classes. For example, using the simple bag-of-words (BOW) approach, sentences are represented as unordered collections of their unigrams and the learning method determines which unigrams tend to occur with each class.

bag-of-words

One can go beyond the BOW approach to learn much more complex patterns. Riloff and Phillips [2004] presented a method for learning subjective extraction patterns from a large amount of data, which takes subjective and non-subjective text as input, and outputs significant lexico-syntactic patterns, that can discriminate between subjective and non-subjective sentences. These patterns are based on shallow syntactic structure output by the Sundance dependency parser [Riloff and Phillips, 2004]. They are extracted by exhaustively applying syntactic templates such as $< subj >$ *passive-verb* and *active-verb* $< dobj >$ to a training corpus, with an extracted pattern for every instantiation of the syntactic template. These patterns are scored according to the probability of a sentence to be subjective given the pattern and the frequency of the pattern. Because these patterns are based on syntactic structure, they can represent subjective expressions that are not fixed word sequences and would therefore be missed by a simple n-gram approach.

The disadvantage of a lexicon-based system is that it usually relies on a hand-built dictionary, which requires many human hours and limits portability to new domains and modalities since vocabularies may differ. An advantage of a statistical system, in contrast, is that porting it to a new domain only requires the new dataset and any requisite annotations from which to learn. The annotation itself may admittedly be time-consuming, depending how coarse or fine it is, but once complete, the system will automatically learn new subjective and polar terms for that domain. On the other hand, the *advantage* of a lexicon-based system is that it can have very high precision, since the dictionaries are typically hand-built and tuned for a particular domain.

As mentioned, the supervised vs. unsupervised distinction is roughly related to the lexicon-based vs. statistical-based distinction in practice. Lexicon-based approaches are often unsupervised, rule-based algorithms (e.g., SO-Cal), while statistical systems typically are trained on labeled data. However, systems can easily cut across these distinctions, e.g., by using the output of a lexicon-based system as a feature of a statistical classifier. And a lexicon-based system itself need not be entirely hand-crafted, but can incorporate words and associated scores that are learned from data in a supervised or semi-supervised fashion.

3.3.2 SENTIMENT DETECTION IN CONVERSATIONS

With meetings, most recent sentiment detection work has focused on the AMI corpus (see Chapter 2). Somasundaran et al. [2007] describe their coding scheme for opinion annotation and apply it to a subset of the AMI corpus. They consider two types of opinions: expressing sentiment, which includes feelings and emotions, and arguing, which includes convictions and persuasion. Their system for detecting sentiment and arguing is a good example of combining lexicon-based and statistical approaches, as they avail themselves of existing sentiment lexicons and create a new arguing lexicon, but also combine these knowledge sources with dialogue act and adjacency pair information in a statistical classifier. Their best results, on both the sentiment and arguing classification tasks, are found by using the basic BOW approach combined with the lexicons and the dialogue information.

Also on the AMI corpus, Raaijmakers et al. [2008] approached the problem of detecting subjectivity in meeting speech by using a variety of multi-modal features such as prosodic features, word n-grams, character n-grams and phoneme n-grams. For subjectivity detection, they found that a combination of all features was best, while prosodic features were less useful for discriminating between positive and negative utterances. They found character n-grams to be particularly useful.

Murray and Carenini [2010] address the same tasks of subjectivity detection and polarity classification as Raaijmakers et al., but on both the AMI corpus and BC3 email corpus. Because they are interested in both spoken and written conversations, their system does not exploit prosodic features as the system of Raiijmakers et al. does, but they nonetheless achieve comparable performance on the AMI corpus. In addition to fixed-sequence n-grams, the authors also introduce *varying instantiation n-grams*, where each unit of the n-gram can either be a word of a word's part-of-speech tag, and make use of lexico-syntactic patterns output by the Riloff and Phillips [2004] algorithm. One finding is that detecting negative polarity sentences is much more difficult than the other sentiment detection tasks, owing partly to the fact that these sentences are relatively rare and can be manifested very subtly. This is particularly true of face-to-face meetings such as the AMI corpus, where negative sentences are not common and seem rarely to be signaled by overt lexical cues.

subjectivity-based summarization

Carenini et al. [2008] are not only interested in detecting subjectivity in emails, but in exploiting that subjectivity information to aid an email summarization system. They take a lexicon-based approach to detecting subjective words and phrases, using existing sentiment dictionaries [Kim and Hovy, 2005, Wilson et al., 2005] and combining measures of subjectivity with measures of lexical cohesion to obtain their best results. Another email summarization system is that of Wan and McKeown [2004], who do not specifically model sentiment but do attempt to summarize discussions that are decision-based, featuring agreements and disagreements, and they have annotated their email corpus for such phenomena, to be exploited in future work. Email summarization systems are described in much more detail in Chapter 4.

There has been a great deal of sentiment and opinion mining research focused on blogs, albeit at a very large-scale, coarse-granularity level. Some of this research attempts to capture a snapshot of the overall blogosphere mood; for example, Mishne and de Rijke [2006] analyze over 8 million LiveJournal[7] posts in order to capture a "blogosphere state-of-mind". The authors learn textual sentiment features by taking advantage of the fact that, in their corpus, many bloggers indicate their mood at the time of each blog post, and the data can therefore be treated as labeled. Mishne and Glance [2006] try to predict movie sales by analyzing the sentiment of blog posts mentioning the movie, and found that considering sentiment improved results over a baseline that only analyzed the volume of postings.

Much of the work on blog opinion mining has emerged under the umbrella of the Text Retrieval Conference (TREC, now the Text Analysis Conference (TAC)). Beginning in 2006, TREC

[7] `http://www.livejournal.com/`

featured a blog "track" consisting of two tasks, one of which was opinion finding. In this task, systems must identify blog posts expressing an opinion about a particular entity, e.g., a celebrity or a company. In TAC 2008, this task was extended to a summarization challenge, so that systems needed to summarize the opinions being expressed.

document re-ranking

In their overview of the TREC Blog Track, Ounis et al. [2008] note that most systems approach the opinion finding task as a two-stage process, where the documents are first ranked using standard information retrieval metrics and then *re-ranked* according to opinion features. They also observe that most systems either automatically build a sentiment dictionary by considering the distribution of terms in opinionated vs. non-opinionated text, or else use a pre-compiled lexicon. In either case, the systems perform better than a baseline that does not consider sentiment features.

The blog sentiment research mentioned so far is not truly conversational, or is only conversational in the loosest sense of millions of people talking about similar topics. They are not talking directly to one another. For example, in the Mishne and Glance [2006] dataset, bloggers might mention the same movie, and in the TREC data many bloggers might mention the same company, but the documents are individual, isolated posts rather than turns within a conversation. In this book, we are primarily interested in conversations where people are responding directly to one another.

Zhou and Hovy [2006] aim to create summaries of such discussions on political blogs. While they do not directly model sentiment within the discussions, they do create separate summaries based on the content of linked articles, thereby generating "factual" news summaries to accompany the opinion-filled blog discussions. Mullen and Malouf [2006] are also interested in political blog discussions, and their goal is to discriminate conservative and liberal users based on the type of language they use in their posts. This work is sentiment analysis in the sense that users are being grouped together according to the shared opinions and emotions that define a political group. The authors found it surprisingly difficult to achieve success on this task using simple textual features, with their 60% accuracy modestly beating the lower-bound.

3.4 CONVERSATIONAL STRUCTURE

In the previous two sections, we have examined topic modeling and sentiment, which are two tasks that can be performed on any document.

In contrast, in this section we will focus on text mining tasks that are specific to conversations. As a start, we discuss what makes conversations unique and so different from other documents.

3.4.1 UNIQUE FEATURES OF HUMAN CONVERSATIONS

turn

A conversation is a joint activity in which two or more participants talk or write to each other. Each contribution to a conversation is called a *turn*, so a full conversation can be described as a sequence of consecutive turns. Spoken, face to face conversations, are by necessity *synchronous*, in the sense that turns must occur one after the other, with minimal delay and minimal overlap. On the contrary, written conversations can be either synchronous like instant messaging or *asynchronous* like email. For asynchronous conversations, consecutive turns can be far apart in time, and multiple turns can largely overlap. For instance, it may be acceptable for an email to be answered days later and for different replies to the same email to be generated in parallel. As a result, while the structure of synchronous conversation can be expected by and large to be linear, asynchronous conversations often display a more complex structure, which, as we will see, can be made even more intricate by the use of quotation.

dialogue acts

Turns in a conversation can perform very different communicative actions, called *dialogue/speech acts* in the literature. For instance, one turn can ask a question, another can provide an answer, and another one can make a request. Notice, however, that a turn can perform more than one dialogue act, especially in written conversations, where for instance an email can ask several questions and request different participants to do different things (see Figure 3.8 for an example).

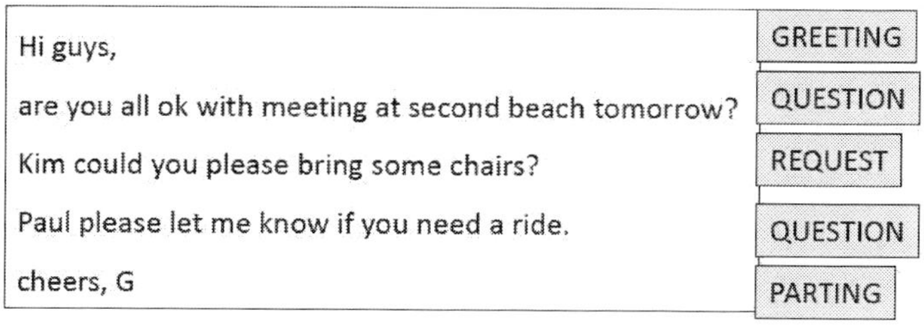

Figure 3.8: Sample email performing multiple dialogue acts. Notice that the fourth dialogue act is formulated as a request, but because it is a request for information it is more appropriately labeled as a question.

There is no commonly accepted standard classification for dialogue acts. Depending on the application, dialogue act tagsets can make very coarse distinctions, like the one shown in Table 3.2 [Jeong et al., 2009], or can be very fine-grained, including dozens of domain specific dialogue acts. For instance, in Vermobil, a machine translation system intended to support people scheduling meetings, the dialogue act tagset comprises 43 dialogue acts, including domain-specific ones like accept-date and reject-date [Jekat et al., 1995].

adjacency pairs

Because of their complementary functions, dialogue acts tend to occur in pairs, called *adjacency pairs*, where the first turn from one participant is generating the following turn by another participant. Common examples of adjacency pairs are, for instance, a question followed by an answer and a request followed by and acceptance or a rejection.

To summarize, conversations have unique properties that clearly distinguish them from monologues. All participants involved in a conversation engage in a joint activity by uttering different types of performative dialogue acts. And these acts are connected to each other, sequentially in synchronous conversations, and in more complicated ways in asynchronous ones.

Mining all this conversation-specific information involves at least the following sub-tasks: recognizing what dialogue act(s) are performed by each turn in the conversation, connecting them to form adjacency pairs, and reconstructing the possibly complex structure of the conversation.

Table 3.2: Dialogue act tag categories.

Tag	Description
S	Statement
P	Polite mechanism
QY	Yes-no question
AC	Action motivator
QW	Wh-question
A	Accept response
QO	Open-ended question
AA	Acknowledge and appreciate
QR	Or/or-clause question
R	Reject response
U	Uncertain response
QH	Rhetorical question

We will also see that, when the focus of the conversation is to make a joint decision and come up with a set of action items, a critical mining task is to detect what turns in a conversation relate to the underlying decision making process. Making joint decisions is much more often the goal for meeting and email conversations, than for blogs, discussion forums and chats.

3.4.2 DIALOGUE ACT MODELING

The task of labeling each turn in a conversation, with the dialogue act(s) it is intended to perform, can be framed as a supervised machine learning classification problem. Since a turn can perform multiple speech acts, a relatively simple technique is to define, for each dialogue act in the tagset, a binary classifier that can determine if a given turn is or is not performing the corresponding dialogue act. Then, to determine the dialogue act labels for a turn, one simply applies all the binary classifiers to that turn and collects the accepted dialogue act labels.

Cohen et al. [2004] have followed this approach, focusing on email conversation in work environments, when people negotiate and coordinate joint activities (e.g., scheduling a meeting). After analyzing several email corpora, they developed an email act tagset which aimed to capture common communication patterns in email usage at work. Their dialogue acts tagset consists of several *verbs* that can be applied to *nouns*; for instance, the act of *delivering* a *Powerpoint presentation* or the act *requesting* the recipient to perform some activity (e.g., *committee membership*). In order to implement and test the supervised approach, the corpora were annotated with this tagset. Agreement among annotator was moderate (κ in the 0.72–0.83 range), which is quite common for dialogue act annotation, especially when the tag are not too specific. Several experiments were then run to compare different feature sets. In general, the performance of the approach is overall not satisfying, with an

F-score for the different dialogue acts ranging from .44–.85. As for features used by the classifiers, the best performance was achieved with a rich set of features, which included features based on the identification of time and date expressions, part of speech and bigrams. For a clear illustration of why bigrams would help in the task, consider the bigrams *"I will"* and *"will you"*. While these two bigrams would strongly indicate a commitment and a request dialogue act, respectively, the three constituent words, *"I"*, *"you"*, *"will"*, in isolation, would be much less informative.

A key limitation of Cohen et al's proposal is that it does not exploit the tendency of dialogue acts to occur in adjacency pairs. It blindly classifies one email message at the time, without considering dependency between a message and its neighbor messages in the email thread. The same research group addressed this limitation the following year in Carvalho and Cohen [2005], where they present an iterative collective classification algorithm[8] in which two classifiers are trained for each dialogue act d_i. One classifier, $Content_{d_i}$, only looks at the content of the message (it is the same classifier presented in Cohen et al. [2004]), whereas the other classifier, $Context_{d_i}$, takes into account both the content of the message and the context in which the message occurs, i.e., the dialogue act labels of its parent and children. The algorithm works as follows.

1. **Initialize** the labels of each message by applying the $Content$ classifiers (which do not need labels for the other messages).

2. **Repeat** for a given number of iterations (60 in the proposal).

    - Revise the labels of all the messages by applying to each message all the $Context$ classifiers.

Figure 3.9 illustrates the algorithm's key operations.

Experimental results show that taking the context into account does improve performance. However, improvements are modest and only for some of the dialogue acts, which indicates that exclusively supervised approaches to email dialogue act labeling may not be the ideal solution.

Similar results are obtained by Shrestha and McKeown [2004], who propose a supervised approach for a rather different dialogue act labeling task. Instead of labeling each message in an email thread with a subset of the labels in a tagset, they only determine whether any two sentences in the thread form a question-answer adjacency pair. On the one hand, this is a more complex task, because it operates at a finer level of granularity (single sentences vs. whole messages), but on the other hand, it is a simpler task because it is limited to identifying only two dialogue acts.

In their work, the detection of question-answer pairs is broken down into two steps. First, you need to identify all the questions in the thread. Next, for each question you need to detect the corresponding answers. Let us examine these two steps in order.

On the surface, it may appear that determining whether a sentence is a question or not in written conversations, like email, should be straightforward, because of the use of the question mark. However, Shrestha and McKeown [2004] discuss three reasons why relying on question marks is not sufficient.

[8]This algorithm is an implementation of a Dependency Network [Heckerman et al., 2001].

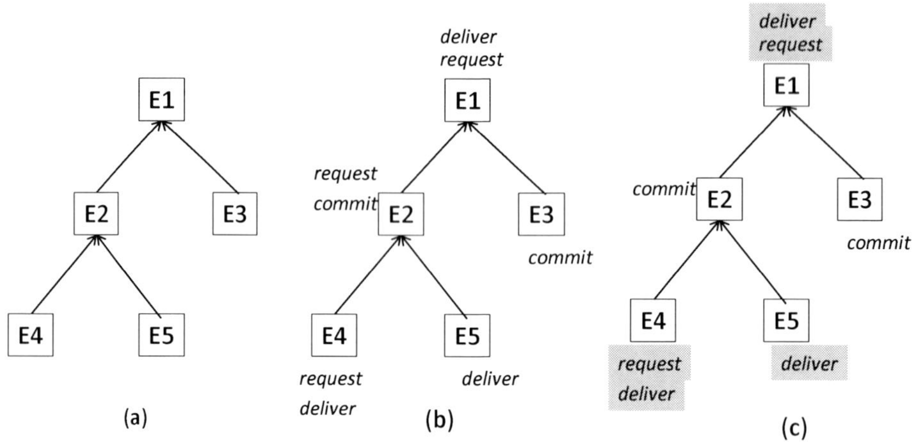

Figure 3.9: Key operations of the collective classification algorithm, assuming a restricted tagset with only three tags, *request*, *commit* and *deliver*. (a) Email thread with five emails initiated by email E1. (b) Step1 - labels for each email message are initialized by applying to each message the three classifiers $Content_{request}$, $Content_{commit}$ and $Content_{deliver}$. (c) As part of one iteration in Step2, the labels of E2 are revised by applying the three classifiers $Context_{request}$, $Context_{commit}$ and $Context_{deliver}$. The contextual information taken into account by the three classifiers is grayed in the Figure.

- Question marks are not always intended to signal a question. Due to the informality of email, and because people strive for conciseness, sometimes question marks are used to express uncertainty.

- Some questions can be phrased declaratively, e.g., "I am wondering if......".

- Some questions are rhetorical in nature; they are not intended to be answered.

Based on these observations, a more sophisticated approach to question detection is proposed (but not thoroughly tested) by Shrestha and McKeown. The task is framed as a binary classification problem, where each utterance is classified as being or not-being a question. In their very preliminary experiments, they only consider a few sentence features, including the length of the sentence and simple lexical features (i.e., POS tags of the constituent words and bigrams). In future work, this set could be expanded with additional lexical and syntactic features.

Once all the questions have been detected, one needs to identify the corresponding answers for each question. This task can also be rather challenging in email. Due to the asynchronous nature of email and the use of quotation, an answer to a question may not appear in the email that directly replies to the email in which the question was originally posed. For instance, in the sample email thread in Figure 3.10, the answer A_{21} to question Q_2 appears in Email-2, a direct reply to the initial email posing the two questions Q_1 and Q_2. However, this is not the case for the answer A_{22}, which appears in Email-5, after two emails, Email-3 and Email-4.

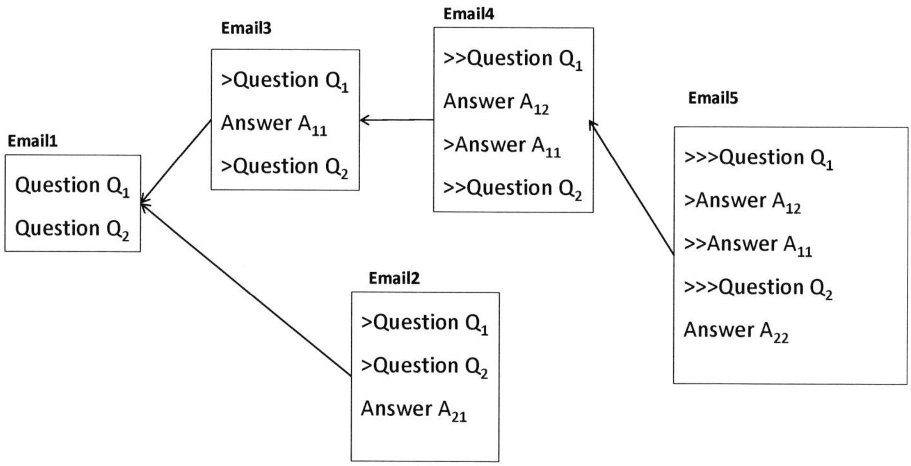

Figure 3.10: Sample email thread that starts with Email-1 which contains two questions, Q_1 and Q_2. These questions receive multiple answers in the following four emails. An answer labeled $A_{i,j}$ means answer j to question i.

For the answer detection task, Shrestha and McKeown also propose a supervised classification approach, where a binary classifier, given a question q, can determine for any utterance u_i following q in the thread, whether or not u_i is a response to q. Even by using a large and complex set of features, based on the lexical similarity between q and u_i as well as the position of q and u_i in the thread, the performance of this approach is modest (F-scores are in the 0.5-0.7 range for different training data).

One critical limitation of this work is that it does not consider quotation as a source of information. As we will see in Section 3.4.5, quotation can be effectively exploited to create a finer-level representation of the conversational structure, which, we will argue, can simplify several mining task, including the dialogue act labeling one. For instance, looking again at Figure 3.10, the answer A_{22} is far from Email-1 (which posed the corresponding question Q_2), but it is adjacent to the quotation of Q_2 (in Email-5).

Although the supervised methods we have discussed so far have generated very useful insights on the task of dialogue act labeling of text conversations, they do require large amounts of annotated data for training, which is not only difficult and extremely time-consuming to build, but also needs to be created for any new conversational modality. By comparison, semi-supervised methods represent a valid alternative, since they can be easily applied to a new domain, as long as you have a considerable amount of unlabeled data in that domain.

Jeong et al. [2009] recently applied a semi-supervised learning method [Bennett et al., 2002] to dialogue act labeling for both email and forum conversations. Their focus is on labeling at the sentence level, in which each sentence is labeled with one of the twelve domain-independent tags shown in Figure 3.2. With respect to previous work (e.g., [Shrestha and McKeown, 2004]), they use a more sophisticated set of sentence features, which includes subtrees of the dependency tree of the sentence [Kübler et al., 2009].

Their semi-supervised approach is essentially an attempt to learn from a combination of labeled transcripts of speech conversations with unlabeled email and forum conversations. For training, they used two large corpora of transcribed spoken conversations as labeled data, namely, a corpus of phone conversations (the SWITCHBOARD corpus), along with a corpus of transcribed meetings (the MRDA corpus). As email unlabeled data, they used a subset of 23,391 emails from the Enron Corpus (see Chapter 2), while as unlabeled forum data they collected 11,602 threads and 55,743 posts from the TripAdvisor travel forum site.

For testing, they annotated with dialog acts all the emails in the BC3 Corpus (see Chapter 2), as well as a small portion of the TripAdvisor posts.

Their experiments reveal several interesting findings. First, more sophisticated sentence features are beneficial for dialogue act labeling. Second, the application of the semi-supervised method was successful, as for both emails and forums the semi-supervised method outperforms a supervised approach in which you simply train on the SWITCHBOARD and MRDA corpora. Third, a closer analysis of the results indicate that the semi-supervised method achieves larger improvements on the less frequent dialogue acts, which suggests that the semi-supervised method is more effective when minimal amount of labeled data are available. Finally, in terms of differences between email and forum conversations, forum data seem to be more challenging, possibly because anyone can post on a forum and this entails more diversity in linguistic and communicative behaviors.

Even more recent work by Ritter et al. [2010] is investigating a completely unsupervised approach to dialogue act modeling, which could be easily applied across new forms of media and new domains. The goal here is less ambitious than full dialogue act labeling. Instead of labeling each utterance (or turn), they cluster together utterances (or turns) that play a similar conversational function. The dialogue act label for each cluster would then be determined through other means, which they do not explore in this work, but may include minimal supervision. Preliminary results on micro-blog data (Twitter) indicate that a sequential HMM-like model can be effectively learned from the data, and that such model reveals interesting properties of the structure of Twitter conversations. For instance, when the states of the model are given some meaningful labels by a human annotator, and transition probabilities of the HMM-like model are visualized as a graph (see Figure 3.11), it becomes clear that Twitter conversations typically start in three different ways: Status, Reference Broadcast and Question to Followers, where a Status dialogue act is describing what the user is doing, a Reference Broadcast act sends out an interesting link or post, and a Question to Followers is self-explanatory. As shown in Figure 3.11, each of these acts can then be followed by different combinations of other dialogue acts with different probabilities. For instance, Status can be followed

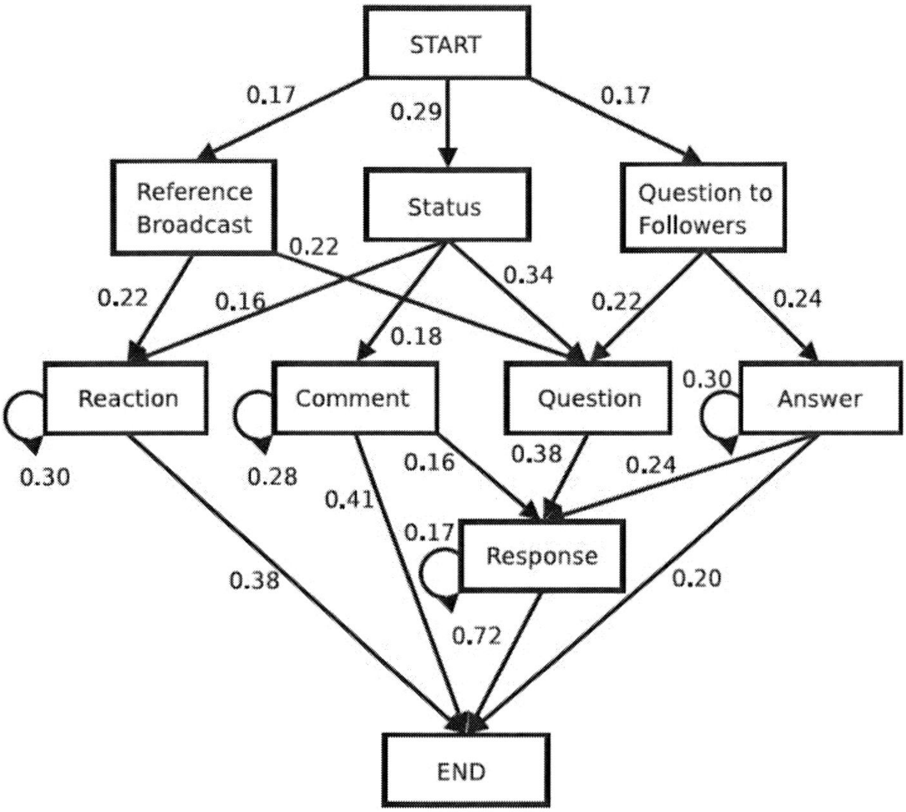

Figure 3.11: Graphical representation of the HMM-like model learned from Twitter data in an unsupervised way. The states were labeled by a human annotator. Transition between states are shown only if their probability is greater than 0.15 [source [Ritter et al., 2010]].

by a Reaction (p=0.16), a Comment (p=0.18), or a Question (p=0.34), but typically not by an Answer (p<0.15)(which is why it is not shown in the graph).

Current and Future Trends in Dialogue Act Modeling for Text Conversations Current research in dialogue act modeling is pushing for semi-supervised and unsupervised approaches and for more sophisticated sentence features. As we already mentioned, another open area for fruitful research is how to integrate the various mining tasks covered in this chapter. For example, an interesting open question, already partially explored in Ritter et al. [2010], is how dialogue act modeling could benefit from topic modeling and vice-versa; or, alternatively, how dialogue act modeling could benefit from extracting a finer grain conversational structure (see Section 3.4.5), as very recently explored in Joty et al. [2011].

3.4.3 DECISION DETECTION

decision process

Related to the problem of dialogue act classification is the task of decision detection. With decision detection, the goal is to identify sentences in a conversation that relate to a *decision-making process*. What separates this task from dialogue act classification is that sentences or utterances may be relevant to a decision without being *performative* decision dialogue acts. Consider the following two sentences:

1. *We need to decide on the type of chip.*

2. *Okay, let's go with the simple chip.*

Both sentences are relevant to a decision-making process, but only Sentence 2 is a performative *Decision* dialogue act in that the utterance "performs the act" of making a decision. The dialogue act type for Sentence 1 would likely be *Statement* or *Inform*, but we still want to capture the fact that it is relevant to a decision process.

Decision sentences are more common in meeting and email conversations than in blogs and discussion fora, since the latter venues tend to be informal and less goal-oriented. Decision sentences are particularly frequent in meetings, since meetings tend to feature a cohesive group of participants working on some type of joint venture. This is not necessarily the case even with email conversations.

Another reason that meetings have been the focus of much decision detection research is the presence of relevant annotations for the AMI and ICSI corpora. As described in Chapter 2, the abstract summaries for these meetings contain subsections for describing decisions made during the meeting. Since there are also abstract-extract links, we can determine which utterances in the meeting relate to decision processes, thereby providing labeled data for supervised decision detection. Figure 3.12 shows example links between utterances from an AMI meeting and the decision subsection of the gold-standard abstract.

An example of such a supervised approach is the work of Hsueh and Moore [2007] and Hsueh et al. [2007]. The authors train maximum entropy classifiers on this labeled AMI data, using lexical, prosodic, topical and contextual features. They also predict decision segments at two levels of granularity: utterance level and topic segment level. They found that this full set of features yielded the best precision but suffered from lower recall than a simple baseline using only unigram features.

issue
resolution
agreement

Fernandez et al. [2008] also worked on decision detection in the AMI corpus, but at a finer-grained level. Rather than just classifying utterances as relevant to a decision process, they created a hierarchy of decision types. The main three types are *issue*, *resolution* and *agreement*. Issue utterances are those that introduce or describe the topic to be decided on. Resolution utterances are those that contain the final adopted decision. There are two subclasses of resolution utterances; *proposal* utterances propose the adopted decision, while *restatement* utterances confirm or restate the adopted decision. Agreement utterances are those that signal agreement with the adopted decision.

Again considering Figure 3.12, we could consider S1 to be an *issue* sentence since it introduces the topic to be decided on (whether they should continue with production), S2 to be a *resolution*

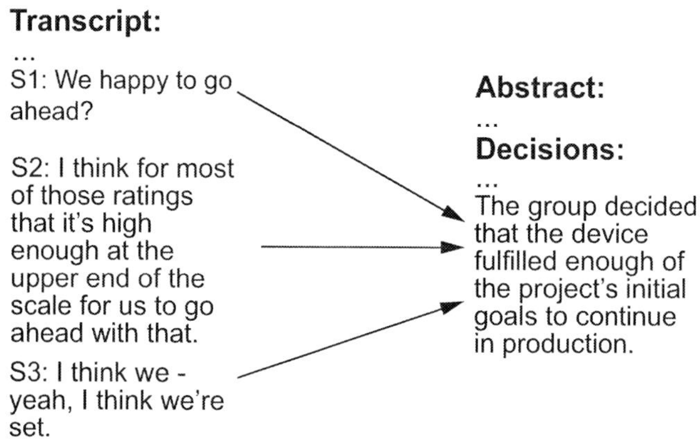

Figure 3.12: Example decision sentences.

(the ratings justify continuing) and S3 to be an *agreement*. The point of resolution may, in fact, be difficult to detect. Only if the speaker of S3 is the project manager who has final say on the matter, then S3 should be considered the resolution of the decision process.

Using similar features as Hsueh and Moore, Fernandez et al. employed SVM classifiers in a two-step approach, where sub-classifiers first predicted each decision class and a super-classifier then took those predictions as its own features. They demonstrated that their hierarchical approach to decision detection achieved a higher F-score than a "flat" approach to predicting the decision classes.

3.4.4 ACTION ITEM DETECTION

Action items are sentences in a conversation that concern the assignment of responsibility for completing tasks. Such responsibilities can be self-assigned or delegated from one person to another, and in this sense are related to dialogue act types such as *commit* and *request*. From the AMI corpus, we can see two examples of the project manager assigning action items to the other meeting participants:

1. *So you will have - Baba and David Jordan you will have to work together on the prototype.*

2. *And you will have next time to show us modelling a clay remote control.*

Note that Sentence 2 contains a particular time-frame: the clay model must be shown by the next meeting. This is typical of action items in meetings, where the action items specify tasks to be done in-between meetings. With emails, an action item may involve one person asking another to do some task and then e-mail back as soon as possible. **time-frame**

The action items described in Sentences 1 and 2 also have "owners" - the person or persons **owners** responsible for the task. In this case, Baba and David Jordan are the owners (the *you* in Sentence 2 refers back to Baba and David Jordan in Sentence 1).

As with decision sentences, action items are more common in meetings and emails than in blogs and discussion fora, for the same reasons. Here, we describe work on those two types of conversations.

In the preceding section, we described how the AMI and ICSI abstract summaries contain decision subsections, and that one can exploit the abstract-extract links to determine which utterances from the meeting are relevant to decisions. The AMI corpus also contains action subsections, and one can similarly exploit the abstract-extract links to obtain a gold-standard labeling of meeting utterances concerning action items.

Murray and Renals [2008] train logistic regression classifiers on AMI data labeled in such a manner, using prosodic, lexical, structural, length and speaker features. On their test set, the highest AUROC score was 0.93 (out of 1), indicating that these features are very effective for identifying action items utterances. Structural features alone matched the performance achieved when using all features, the simple explanation being that action items are much more likely to occur at the end of a meeting. Lexical features were also effective; note that both Sentences 1 and 2 above are signaled by the pattern *you will*, and such cues are common.

description
agreement

Purver et al. [2006a] take a finer-grained approach to action item detection, using the ICSI and CALO corpora. Rather than just classifying sentences as being related to an action item, they aim to detect subclasses of action item utterances. The four classes are *description*, *time-frame*, *owner* and *agreement*. Owner and time-frame are described above. Description utterances are those that describe the task to be carried out. Agreement utterances are those that indicate acceptance or agreement with an assigned action item. The authors found that this finer-grained classification is quite challenging, with the Agreement class being the easiest to detect of the four (F-score of 0.40), while the owner class being the most difficult (F-score of 0.17). Ownership of action items in meetings is so difficult because the person doing the delegating may not explicitly mention to whom the item is being assigned; rather they will simply speak directly to that person. In that sense, ownership detection is closely related to the *addressing* problem of determining which speakers are addressing one another. In subsequent work, Purver et al. [2007] investigate summarization of action items by extraction of useful phrases such as those including explicit time-frames, and find that this condensing stage can in some cases yield more informative results than simply giving the utterance transcriptions.

Using such fine-grained analysis of action items, we could extract useful, specific information from Sentences 1 and 2 above, as shown in Figure 3.13. We can now present the user with descriptions of the action items, as well as the owners and time-frames, and index the sentences from which these pieces of information were taken.

On email data, Bennett and Carbonell [2005] aim to classify each message according to whether the sender requires a response of some kind, as well as identifying the particular sentences that mention the action items. They employ several types of supervised classifiers and compare

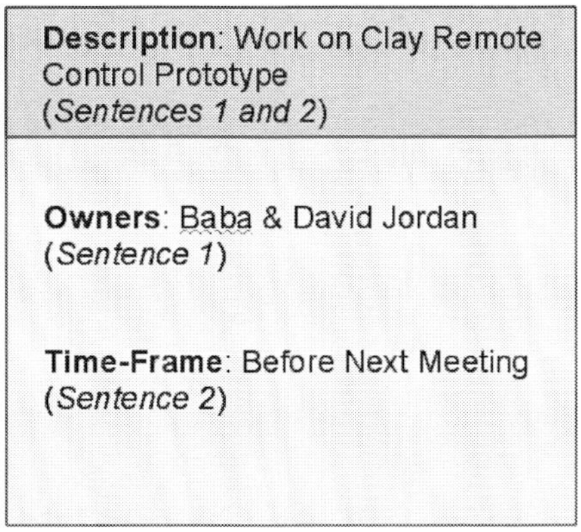

Figure 3.13: Extracting action item information.

unigram features with higher-order n-grams, concluding that higher-order n-grams are superior for these tasks. The best reported F-score for document-level classification is 0.78.

Corston-Oliver et al. [2004] also aim to detect action items in emails in order to create an email "to-do" list, but with a much richer set of features and larger corpus. They train SVM classifiers using three feature types: *message* features such as the number of recipients and the message size, *superficial* features such as n-grams, names and special characters, and *linguistic* features including part-of-speech n-grams and features of logical form. Like Purver et al. [2007], Corston-Oliver et al. subsequently try to summarize the action items in order to create a succinct to-do list. This post-processing step involves reformulating the sentences by identifying the clauses containing the task, deleting extraneous words, replacing certain deictic expressions with non-deictic expressions, and replacing all temporal expressions with absolute dates.

A very recent example of work on detecting action items in email is Lampert et al. [2010], where the authors show that determining whether an email contains a request for action can benefit from an initial segmentation of the email into nine functional zones. These zones include: content written by the current sender, greetings, sign offs, quoted reply content, forwarded content, etc. A competitive F-score of 0.84 is reported for experiments run on an annotated subset of the Enron corpus.

Both action item detection and decision classification can be considered a type of focused summarization, and we discuss summarization in much more detail in Chapter 4. For now, we simply note that decisions and action items are the types of information that many people will desire to know about when they have missed a meeting or want to digest a long email conversation. They want to know what decisions have been made, and what they are required to do.

3.4.5 EXTRACTING THE CONVERSATIONAL STRUCTURE

At first glance, it may seem that the task of extracting the structure of a conversation should be a rather easy one. While for asynchronous conversations, like emails and blogs, the conversational structure should be fully revealed by the reply-to relation between messages; for synchronous conversations, such as meetings and chats, the structure should simply consists of the linear sequence of turns appearing one after the other as the conversation evolves.

However, if you think more carefully, there are two limitations with this initial, simple view of conversational structure. First, in asynchronous conversations the use of quotation can express a conversational structure that is at a finer level of granularity than the one revealed by reply-to relations between emails or blog posts. For instance, as we saw in the email example in Figure 3.10, the proximity between a quoted paragraph and an unquoted one can represent an informative conversational link between the two (i.e., question/answer adjacency pair) that would not appear by only looking at the reply-to relations. Secondly, the linear structure of synchronous conversations can be misleading in its simplicity. An empirical analysis of such conversations, of both meetings and chats, show that what appears to be single, linear conversation, may in fact contain several simultaneous conversations that need to be disentangled.

disentangled

Let us now examine how we can deal with these two additional complexities for mining text conversations. More specifically, how can we extract the finer granularity conversational structure induced by the use of quotation in asynchronous conversations? And, how can we disentangled simultaneous conversations in seemingly single, linear synchronous conversations?

Building the Fragment Quotation Graph Since in asynchronous conversations consecutive turns can be far apart in time, when people reply to an email or comment on a blog post, a quotation of the original message is often included by default in the draft reply in order to preserve context. Furthermore, people tend to break down the quoted message so that different questions, requests or claims can be dealt with separately. If, for instance, the original message is asking multiple questions, the replier might type each answer under the corresponding question. As a result, each message, unless it is at the beginning of a thread, will contain a mix of quoted and novel paragraphs that may well reflect a reply-to relationship between paragraphs (or sentences) that is at a finer level of granularity than the one explicitly recorded between emails.

Carenini et al. [2007] propose a novel approach to capture this finer level conversational structure of asynchronous text conversations in the form of a Fragment Quotation Graph (FQG). We describe the construction of a sample FQG by following an example originally presented in Carenini et al. [2007]

Fragment Quotation Graph

Figure 3.14(a) shows a real example of a conversation from the Enron Corpus involving six emails. For the sake of illustration, we do not show the original text, but abbreviate it as a sequence of labels $< a, b, c, ..., j >$, each one corresponding to a text fragment, typically a sentence or a paragraph. To build a FQG, you follow a two-step process.

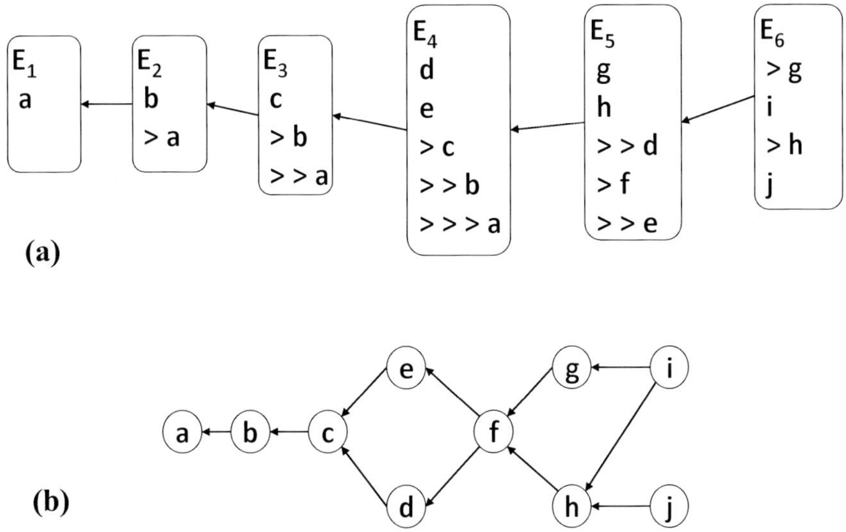

Figure 3.14: (a) An email conversation from the Enron corpus containing six emails. Arrows between emails represent the reply-to relation. (b) The corresponding Fragment Quotation Graph, in which nodes are created by identifying quotations and edges are created between neighboring quotations.

- *Creating Nodes:* Initially, all new and quoted fragments are identified. For instance, email E_2 is split into two fragments: the new fragment b and the quoted fragment a. E_3 is decomposed into 3 fragments: the new fragment c and two quoted fragments b and a. E_3 is decomposed into de, c, b and a, and so on and so forth. After that, to identify distinct fragments (nodes), fragments are compared with each other and overlaps are identified. Fragments are split if necessary (e.g., fragment gh in E_5 is split into g and h when matched with E_6), and duplicates are removed. At the end, 10 distinct fragments a, ..., j give rise to 10 nodes in the graph shown in Figure 3.14(b).[9]

- *Creating Edges:* Edges are created to represent likely replying relationship among fragments. The assumption is that any new fragment in a message is a potential reply to neighboring quotations, i.e., quoted fragments immediately preceding or following it. For instance, consider E_6 in Figure 3.14(a), there are two edges from node i to g and h, because i is between g and h; while there is only a single edge from j to h, because j is under h, but there is no text under j.

Figure 3.14(b) shows the complete fragment quotation graph of the conversation shown in Figure 3.14(a). Notice how the threading of the conversation in the FQG is done at the finer level

[9]In this email thread, fragment f reflects a special and important phenomenon, where the original email of a quotation does not exist in the thread. Carenini et al. characterize this as the hidden email problem and its influence on email summarization is discussed in Carenini et al. [2007].

granularity than entire emails, where the thread would be a simple chain from E_6 to E_5, E_6 to E_4 and so on.

(a)

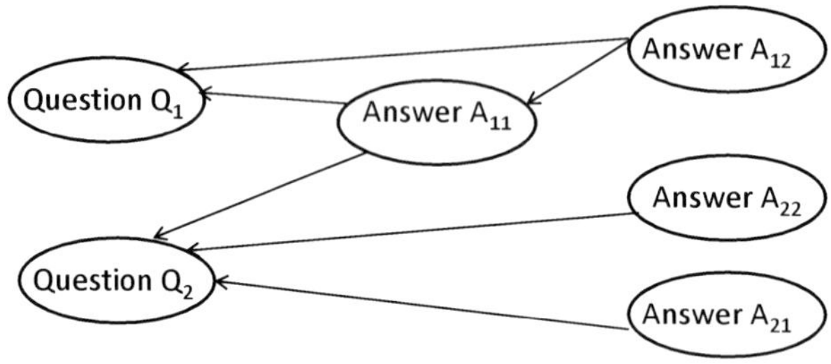

(a)

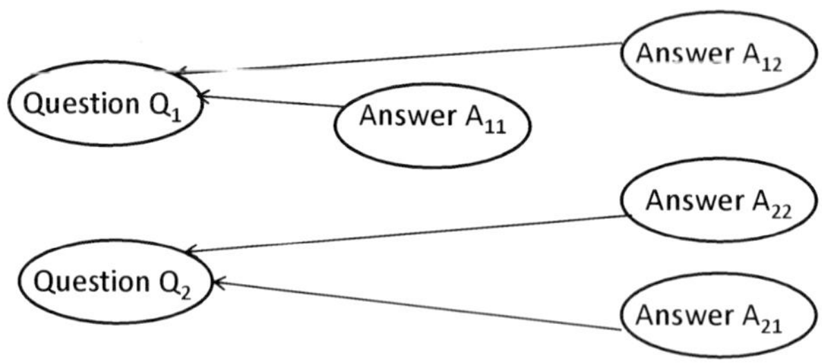

Figure 3.15: (a) FQG for the dialogue act labeling example in Figure 3.10. (b) FQG for the same example when an edge is created only between to fragments if one is below the other in an email.

As mentioned, the FQG is only an approximation of the reply relations between fragments. In some cases, proximity may not indicate any connection and in other cases a connection can exist between fragments that are never adjacent in any email in the thread. Nonetheless, Carenini et al. [2008] showed that considering the FQG can be beneficial in email summarization and we argue that similar benefits could be obtained if the FQG was used to support the other text mining tasks discussed in this chapter. For instance, let us go back to the dialogue act modeling example shown in Figure 3.10. The FQG for that example would be the one shown in Figure 3.15(a). Remarkably, if we had identified the fragment asking the questions, it would be straightforward to identify most of the corresponding answers on the FQG. The only ambiguous case would be A_{11}, which could be an answer to both Q_1 and Q_2. One heuristic to resolve this problem could be to build the FQG in a more conservative way, namely, by creating edges in the FQG between to fragments only if one is below the other in the text, which would result in the FQG in 3.15(b).

Disentangling Synchronous Conversations On the surface, synchronous conversations, such as meetings and chats, consist of the linear sequence of turns appearing one after the other as the conversation evolves over time. However, a single stream of turns often contains several simultaneous conversations. For instance, Aoki et al. [2006] found that in speech conversations involving 8 to 10 speakers an average of 1.76 distinct conversations were occurring at the same time. This phenomenon was shown to be even more pronounced in chats, where a recent study has found that an average of 2.75 conversations were simultaneously active [Elsner and Charniak, 2010].

The difference between spoken conversations and chats can be explained by considering one aspect of chats that make them similar to asynchronous text conversations (e.g., email): chats do not allow participants to control the positioning of their contributions [Smith et al., 2000]. In other words, if you send an answer to a question in a chat, since several participants can simultaneously send different messages, there is no guarantee that your answer will follow the original question. Other contributions, possibly unrelated to the question, may appear between the question and your answer.

Since in chats and in other synchronous conversations, what appears to be like a single stream of turns often contains multiple, independent, interwoven conversations, we are faced with the challenge of identifying those conversations; that is, we must disentangle the conversations.

A two-step approach to disentangling conversations has been recently proposed by Elsner and Charniak [2010]. The first step is based on supervised classification. For each pairs of turns (x, y) in a chat stream, a binary classifiers determines how likely is that the two turns x and y belong to the same conversation. The classifier is trained on a set of features that are grouped into three classes (see Elsner and Charniak [2010] for a complete list):

- *Chat-specific:* including, for instance, the temporal distance between x and y and whether x mentions the speaker of y (or vice-versa).

- *Discourse:* including, for instance, whether x and y uses a greeting word ("hello" etc.), an answer word ("yes", "no" etc.), or the word "thanks"; or asks a question (marked explicitly with a question mark)

- *Content:* including, for instance, whether x and y both use technical jargon, neither do, or only one does.

In the second step of the disentangling process, turns are clustered by relying on the output of the classifier used in step one. In essence, the clustering algorithm tries to make sure that pairs of turns likely to belong to the same conversation (according to the classifier), will be in the same cluster, while pairs that are unlikely to belong to the same conversation (again, according to the classifier) should end up in different clusters. The goal is that each resulting cluster will correspond to a different conversation.

Unfortunately, finding the optimal solution to this clustering problem is intractable. However, the authors show that acceptable solutions can be found with heuristic algorithms.

With respect to the other mining tasks, disentangling conversations should be performed first, otherwise topic modeling and dialogue act modeling would be "confused" by the mixture of independent conversations. Only once a conversation has been disentangled should each extracted conversation be processed separately.

Current and Future Trends in Extracting the Conversational Structure

- A Fragment Quotation graph, in which links are added only based on proximity, may still represent a too rough approximation of the reply-to relation between fragments, as proximity can be misleading and the granularity of the fragments can be incorrect. So an open question is how we can build a FQG that better reflects the true structure of the conversation. Again, we believe that integrating the construction of the FQG with the other mining tasks (e.g., topic modeling) can be a key part of the answer.

- According to Elsner and Charniak [2010], it is the first step of disentangling conversations, namely supervised classification, where improvements can be more easily obtained. On the one hand, better classifiers could be deployed. On the other hand, a richer set of features, based on lexical coherence, could be applied. Moreover, as it is the case for the other mining tasks, we believe that the application of completely unsupervised methods should be further explored.

3.5 CONCLUSION

In this chapter, we considered several mining tasks relevant to conversations, including sentiment and subjectivity analysis, topic segmentation and modeling, extraction of thread structure, dialogue act classification, and detection of decisions and action items. Some of these tasks can depend on others; for example, work on action item detection can use predicted dialogue act labels as an input feature to a statistical classifier.

In Chapter 4, we will see that the outputs of these mining systems can also be used as inputs to a summarization system. In fact, many of these mining systems already constitute summarization systems of a particular kind. Subjectivity systems summarize the opinions and sentiment in a conversation, topic systems identify and describe the topics discussed, while action item and decision detection condense the conversation down to critical bullet-points.

3.6 IMPORTANT POINTS

- Topic segmentation is the task of splitting an input document into multiple segments, where each segment corresponds to a single topic.

- Topic modeling consists of topic segmentation plus topic labeling, where each topic in the document is given an informative description, such as a set of words.

- Research on detecting opinions, sentiment and subjectivity is concerned with identifying what people think or feel, in terms of opinions and emotions expressed.
- A dialogue/speech act represents the illocutionary meaning of an utterance, or the action performed by the utterance. Dialogue act modeling is the task of labeling each conversation turn with the dialogue act(s) it is intended to perform.
- Decision detection is the task of identifying sentences related to a decision process. This may involve identifying sub-types of decision sentences.
- Action item detection is the task of identifying sentences that relate to the assignment of responsibility for completing tasks. This may also involve identifying sub-types of action item sentences.
- Decision and action item detection can be thought of as focused summarization.
- Synchronous conversations, especially written ones like chats, often require disentanglement to determine reply-to relationships between turns.
- Extracting a fragment quotation graph from a conversation can reveal a finer level conversational structure which can be beneficial to other mining tasks.

3.7 FURTHER READING

For subjectivity research, the best general reference is by Pang and Lee [2008]. They include numerous case studies including research on blog data. The book is available online[10] and the book site contains a useful searchable bibliography.

Jurafsky and Martin [2008] have a chapter on dialogue and conversation, including coverage of dialogue acts and adjacency pairs, as well as a discussion of the relation between dialog acts and the theory of speech acts [Austin, 1962], [Searle, 1975]. This book also includes discussions on many mining tasks not covered here, such as named entity recognition, relation extraction and rhetorical parsing.

On topic segmentation, Jurafsky and Martin briefly describe unsupervised and supervised approaches as well as segmentation evaluation. A very recent, comprehensive review on topic segmentation is provided by Purver [2011].

On topic modeling, David Mimno has maintained a bibliography of relevant papers and available software[11]. A review paper has also been posted by Blei and Lafferty[12], and Steyvers and Griffiths [2006] give a very gentle introduction to probabilistic topic models such as LDA.

[10] `http://www.cs.cornell.edu/home/llee/opinion-mining-sentiment-analysis-survey.html`
[11] `http://www.cs.princeton.edu/~mimno/topics.html`
[12] `http://www.cs.princeton.edu/~blei/papers/BleiLafferty2009.pdf`

In this chapter, we assumed that for asynchronous conversations the reply-to relations between turns are readily available. However, when this is not the case, methods have been proposed to reconstruct the missing thread structure [Wang et al., 2008].

Another interesting topic we do not cover in this chapter is how mining text conversations can benefit from an analysis of the participants' social network. See McCallum et al. [2007] for an example of how this can be done for topic modeling of email conversations. See also the *Synthesis Lecture* by Tang and Liu [2010] for more information on community detection and mining in social media.

CHAPTER 4

Summarizing Text Conversations

4.1 INTRODUCTION

In this chapter we discuss automatic summarization techniques for conversational data. Owing to their prevalence and to the availability of several annotated corpora, emails and meetings are the conversational domains that have received the most attention in the summarization research community. We first discuss some of the approaches taken for these two conversation types, before proceeding to discuss work with online chats, blogs and forums. As we will see, many of the summarization approaches are domain-specific due to the features used, such as email header information or blog comment ratings. However, we conclude with a brief survey of summarization approaches designed for conversations in any modality and domain, or indeed for conversations that span modalities and domains. This includes a detailed case study of an abstractive summarization system for conversations.

In each section, we first present individual examples of summarization systems for the given domain and subsequently use those systems as an entry-point for discussing the assumptions, requirements, inputs and outputs of possible summarization systems in that domain.

To motivate the task of conversation summarization, let us consider an actual conversation. This conversation is taken from the Enron email corpus:

- ***From Erica:*** *Dear Mr. Skilling: I write on behalf of Jerry Murdock, who is currently in Europe. Mr. Murdock asked me to pass on the following information regarding your telephone call scheduled for 10:00am (CST) on Tuesday, May 15. Insight Capital hosts a quarterly dinner with Robert Rubin and Steve Friedman and other selected guests. Our next dinner is scheduled for July 19 and will be held in Aspen, Colorado. The Current State of the Global Market will be one of the topics under discussion and Jack Welch of GE is one of your fellow invitees. The purpose of your conversation with Mr. Murdock is to discuss the above in more detail and to more fully brief you on the purpose of these dinners. Please do not hesitate to contact me should you require any further information.*

- ***From Joannie:*** *Erica, Jeff is currently scheduled to be on vacation July 19. Would it be possible to schedule during the next quarter?*

- ***From Erica:*** *Probably the best thing is for the call to go ahead, that way Jerry can brief Mr. Skilling on our plans for the remaining dinners throughout the year and Mr. Skilling can decide on the best one for him to attend. Does that sound feasible to you?*

- ***From Joannie:*** *Erica, Due to the fact that Jeff is unable to attend on July 19, I believe it would be better to reschedule the call for sometime next quarter.*

- ***From Erica:*** *Joannie: As I mentioned, Jerry is in transit in Europe at the moment, and it will be extremely difficult for me to get hold of him again today to reschedule. Given the time difference involved I will not be able to contact him before the appointed time tomorrow, and therefore I'd very much appreciate if we could go ahead with the call as planned. At least that way Mr. Skilling can decide which of the dinners he would prefer to attend, I assume that the more notice you have the better.*

- ***From Sherri:*** *Erica, thanks for the note. Joannie has left the office for the day, but will return tomorrow morning. In the meantime, I'll run this by Jeff to see if he thinks it makes sense. Do you have any dates for future dinners in mind? The vast majority of Jeff's time is committed through February 2002, so knowing what the dates/timeframes are would be most helpful in the event we need to try to free up some time. Thank you.*

- ***From Erica:*** *Sheri: Thanks for the response. Our next dinner is scheduled for September 25, at the moment I'm not sure who the other invitees would be. Jerry would probably have a better idea as he puts together the guest list for each event and may already have something in mind. That's actually another reason why I think we should leave the call tomorrow on the calendar. Depending on Mr. Skillings availability/interest Jerry might want to reconfigure his invitee list for the most appropriate mix of people.*

Because we have chosen a relatively brief email for purposes of explication, summarization might not strictly be necessary here. However, other email threads are much longer and feature more complicated thread structures. Before we even consider summarization and the question of salience or importance, there is a variety of dimensions on which we can characterize this conversation. We can enumerate the actual participants in the discussion, in this case Erica, Joannie and Sherri. We can identify a number of topics such as *phone calls*, *travel* and *dinners*. We can extract dates (*July 19*, *September 25*) and named entities (*Europe*, *Aspen*, *Insight Capital*, etc.). We can observe that there is an initial email that contains a great deal of information, and several shorter follow-up emails. We can detect sentiment and opinions, such as Joannie and Erica disagreeing about whether the phone call should go ahead. Essentially, we can enlist all of the mining techniques described in Chapter 3 in order to derive some structure from this conversation. As previously stated, many of those mining techniques themselves can be considered a type of focused summarization, where particular types of information are being extracted, and a large part of a general summarization system may consist of combining those pieces of information.

Given the email and the derived information, how might one summarize this conversation for a third party? The answer partly depends on the audience. The two people who are the focus of the conversation, Mr. Murdock and Mr. Skilling, did not take part in the discussion, and a summary generated for the benefit of Murdock may differ from the summary provided to Mr. Skilling. The summary for Mr. Skilling might detail the time and purpose of the phone call, whereas the summary for Mr. Murdock might simply confirm that Skilling will participate. This demonstrates there is no such thing as a single best summary for a given conversation. There are many other reasons why multiple summaries of a single conversation might differ, such as ambiguity inherent in the conversation. If we ask two people to create summaries of this email exchange, they may disagree on whether a decision was actually made regarding the phone call. Furthermore, summaries can vary according to explicitly provided information needs. A summary might not be generated generically but rather in response to a user query, in which case a good summary could be focused on a particular date, time or name.

Summaries of this conversation might also vary in granularity. We could generate a concise, decision-based summary that reads, *It was decided that Mr. Skilling and Mr. Murdock will speak on the phone tomorrow*, or we could generate a summary that describes the decision *process*, in this case emphasizing that Erica and Joannie disagreed about the need for the phone call. The decision process in this example might seem trivial and unnecessary to summarize, but in many real-world cases it is the decision process that is critical to understand. In fact, automatic summarization has previously been touted for its use in conducting corporate decision audits [Murray et al., 2009]. While an important decision is likely to be well known and disseminated within an organization, the decision *process* might quickly be forgotten. If it turns out that the decision was ill-advised, reconstructing the decision process may be in the interest of the organization, in order to determine responsibility and accountability.

decision audit

One can imagine many ways to summarize this conversation, based on decisions, agendas, action items, opinions or some combination thereof. Keep these possibilities in mind as we discuss specific work in a variety of conversational domains.

4.2 SUMMARIZATION FRAMEWORK AND BACKGROUND

In discussing automatic summarization of conversations, we describe summarization approaches and systems according to three aspects:

- *Assumptions and Inputs.* Assumptions can mean assumptions about the nature and format of the data or assumptions about an end user's information needs, to give just two examples. Inputs can mean upstream modules such as preprocessing and information extraction.

  Assumptions and Inputs

- *Measuring Informativeness.* This describes how a given approach or particular system determines salience for a conversation, and constitutes the heart of the summarization pipeline.

  Measuring Informativeness

- *Outputs and Interfaces.* Outputs can refer to the modality of the summary (e.g., textual vs. visual) and more specifically to the structure of the produced summary (e.g., extractive vs. abstractive text). Interfaces can refer to the manner in which the summary is meant to be used by an end user.

  Outputs and Interfaces

There are some parallels between these aspects and the traditional summarization pipeline of *interpretation*, *transformation* and *generation* [Jones, 1999], although our categories are more general and not necessarily meant to describe individual components or modules in a system. In the following sections we briefly introduce these three aspects, and subsequently describe them in detail when discussing actual summarization systems in each conversation domain.

4.2.1 ASSUMPTIONS AND INPUTS

By *assumptions and inputs*, we include such aspects as the nature of the data to be summarized (e.g., genre, style), the *representation* of the data, and upstream processes on which summarization depends. In Chapter 1, we also provided some basic distinctions within summarization that can be thought of as assumptions about the corpus or about the summarization task itself, e.g., single-document vs. multi-document summarization, extractive vs. abstractive summarization, generic vs. query-based summarization, and informative vs. indicative summaries. Cutting across those distinctions are common representations and features that we can now introduce.

First, we need a way of representing documents and for representing queries in the case of query-based summarization. One popular option is the *vector-space model* in which documents (and queries) are represented as vectors of features and the features represent the words present in the document. Note that document here can mean a single sentence or a collection of sentences. In the simplest case, given a document and a list of all words from the entire document set, we could have a 0/1 binary feature for each word indicating whether or not it occurred in that particular given document. More often, we will represent each word (or term) using a *term-weighting* scheme. The idea behind term-weighting is to weight certain words more highly than others, based on criteria such as word frequency in the document. The most widely used term-weighting scheme is *tf.idf*, where *tf* stands for term-frequency and *idf* stands for inverse document frequency. More precisely, *tf* is given by

vector-space model

term-weighting

tf.idf

$$TF(t,d) = \frac{N(t)}{\sum_{k=1}^{T} N(k)},$$

where $N(t)$ is the number of times the term t occurs in the given document d and $\sum_{k=1}^{T} N_k$ is the total word count for the document, with the denominator normalizing the term count by document length. The idf is given by

$$IDF(t) = -\log(\frac{D(t)}{D}),$$

where D is the total number of documents in the document collection and $D(t)$ is the number of documents that contain the term t. A term will therefore have a high IDF score if it occurs in only a few documents in the collection. The $tf.idf$ measure simply multiplies these two weights. Obviously, the $tf.idf$ scores will depend heavily on how we define the document collection. In the sample conversation above, the word *call* has a high term-frequency in the document (here the conversation is a document). However, if the document collection is a collection of other emails (such as the contents of Mr. Skilling's inbox) and those emails often discuss telephone calls, the idf score may be quite low and the term will not be weighted highly.

It should be clear why term-weighting is relevant to summarization. The goal of summarization is to identify the most important information in a document, and term-weighting is a useful tool for identifying important or significant words in a document. One of the simplest summarization approaches, then, would be to extract the sentences with the highest $tf.idf$ scores (e.g., by summing or averaging over each sentence). Indeed, this can be a surprisingly decent baseline summarizer in some cases. But in the following section, we will see that there exist much more advanced methods of measuring informativeness, and many useful feature types beyond the term-weights described above.

Each of the mining techniques described in Chapter 3 can be considered a potential input to a summarization system. For example, a system may depend on having fine-grained sentiment analysis or decision detection. Many summarizers utilize topic detection or clustering modules. In particular, conversation domains such as meetings and emails, summarization systems may make assumptions about the data and metadata that are available to the summarizer, and we will discuss these in each subsection.

All *conversation* summarization systems share the simple assumptions that the input is a multi-party exchange featuring turn-taking and interactions. Indeed, these characteristics define conversation itself and are the common link between meetings, emails, blogs and discussion forms, and set these domains apart from lectures, broadcast news and articles, all of which feature little or no conversation. Beyond those common characteristics of turn-taking and interaction, conversations can widely differ in terms of number of participants, goal-directedness, synchronicity, etc. A summarization system designed for a particular domain such as meetings might make assumptions about the nature of a conversation in that domain, e.g., that it has a definite beginning and end and that conversation participants have specific roles, and these assumptions may not be true of other conversation domains such as blog comments or discussion forums.

In terms of inputs, conversation summarization systems diverge according to how the conversation is documented. For meetings, there may be a rich, multi-modal corpus of data including transcripts, audio, video, and notes. Email threads contain the email text in addition to metadata from the email header, and possibly attached documents. Blogs contain posts, comments and links to other webpages. Some summarization systems might eschew the multi-modal data and process only the available text of the conversation discussion itself.

In the discussion of each domain below, we will describe what assumptions various summarizers embody when applied to that domain. We will also see that summarization systems that are designed to work on conversations *across* different domains must make comparatively few assumptions about the nature and structure of their inputs.

4.2.2 MEASURING INFORMATIVENESS

In the previous section, we introduced the idea of representing sentences using term-weights and proposed a very simple summarizer that weights sentences by summing or averaging over the constituent term-weights. However, we can use these term-weights to build a considerably more sophisticated summarizer. By representing sentences as *vectors* of term-weights, we can measure the similarity of two sentences by calculating the *cosine* of the angle between their vectors. This similarity metric is essentially the normalized dot product of the vectors and will range from 0 when sentences share no terms to 1 when sentences are identical. We can now use this *cosine similarity* metric in a variety of ways; if we are doing query-based summarization, we can calculate the similarity of a candidate sentence to the query. If we are doing multi-document generic summarization, we can calculate the similarity of a candidate sentence with the set of sentences already selected for extraction. In fact, this is precisely what is done in the popular *Maximal Marginal Relevance* (MMR) [Carbonell and Goldstein, 1998] summarization approach.

cosine similarity

MMR

In MMR, sentences are chosen according to a weighted combination of their relevance to a query (or for generic summaries, their general relevance) and their redundancy with the sentences that have already been extracted. Both relevance and redundancy are measured using cosine similarity. The usual MMR score $Sc_{MMR}(i)$for a given sentence S_i in the document is given by

redundancy

$$Sc_{MMR}(i) = \lambda(cos(S_i, q)) - (1 - \lambda) \max_{S_j \in summ} (cos(S_i, S_j)),$$

where q is the query vector, $summ$ is the set of sentences already extracted, and λ trades off between relevance and redundancy. The term cos is the cosine similarity between two documents. The MMR algorithm iteratively generates the extractive summary, at each step selecting the sentence i that maximizes $Sc_{MMR}(i)$ and recalculating the scores of the remaining unselected sentences. This recalculation is necessary because the redundancy scores will have changed each time a new sentence is added to the summary. However, if λ equals 1 then redundancy scores will be ignored and MMR will return the sentences most similar to the query.

Whereas MMR is an unsupervised extraction algorithm, many recent extractive systems are supervised machine learning approaches and incorporate a variety of features in addition to term-weights such as *tf.idf*. A classifier is trained on data where each sentence is hand labeled as informative or not informative, and sentences in the test data are classified as informative or non-informative based on the trained model. In the sections below we will discuss the types of features used by different supervised systems. Because the supervised classifier is typically only predicting the relevance of the candidate sentences, such summarization systems will often incorporate a post-classification step designed to reduce redundancy. This might involve clustering the informative sentences and selecting only a handful from each cluster.

clustering

This preceding discussion of informativeness is relevant primarily to the extractive paradigm, and the architecture of *abstractive* systems is typically much different. Rather than rating the informativeness of individual sentences, abstractive summarizers tend to look for patterns, messages or events that abstract over numerous sentences. Informativeness might be based at least partly on

messages

knowledge about the domain and which types of events are often significant in that domain. For example, the summarizer could detect that there are many sentences in a meeting or email thread that concern a particular action item, and the awareness that action items often form a critical part of a summary might be part of the knowledge base available to the summarizer.

knowledge base

ontology

More specifically, domain knowledge can be represented in an *ontology*. One popular ontology language is OWL/RDF, widely used in semantic web contexts and based on description logics, a subset of first order logic. An ontology typically contains a class-subclass hierarchy, properties or relations, and instance data. For example, we may have a class *Person* and a subclass *Manager*, and a particular instance of a manager *Heather*. We may have a property or relation *worksWith* that connects instances of *Person*. Adding instance data to the ontology is called *populating the ontology*. To use the language of ontological engineering, our classes and properties are defined in the T-Box of the ontology, while the A-Box contains our instance data. We do not go into any more detail on ontologies here, but any primer on the semantic web should suffice to give a general overview (e.g., [Allemang and Hendler, 2008], [Segaran et al., 2009]).

An abstractive system also requires natural language generation (NLG) to create the summary output. An NLG system is typically comprised of a *planner* to create the document structure, a *micro-planner* to refine the document plan by doing aggregation and coreference resolution among other tasks, and a *realizer* to generate the actual surface text. Reiter and Dale [2000] provide the classic text on NLG systems and components.

4.2.3 OUTPUTS AND INTERFACES

Although research on automatic summarization usually concerns the generation of textual summaries, summaries need not be text-based. A meeting summary could consist of concatenated audio clips from the discussion, while any conversation could be summarized with a graphics-based visualization highlighting information such as participant activity and topic dispersion.

visualization

Even with textual summaries, many types of output are possible. One could generate well-formed paragraphs of coherent text describing the conversation at a high level—a so-called *abstractive* summary necessitating a text generation component. A simpler approach, but one that leads to less coherent summaries, is the *extractive* approach of simply classifying some sentences as important and pasting them together. One could also generate a *word cloud* or a list of dates, named entities or keywords. A word cloud might not seem like a summary in the traditional sense, but it does fit the definition of condensing a document to a simple representation of its most important components. For example, Figure 4.1 shows a word cloud representing the email conversation shown earlier.

word cloud

Summary outputs and interfaces also vary according to how a summary is intended to be used. If a summary is meant to serve as an index to browsing the original document, then it might be situated in a browsing interface along with a variety of other search and browsing functions. Indeed, a browing interface could feature multi-modal summary types such as an abstractive textual summary alongside a word cloud and some visualizations of participant activity. All of these summary types would then be linked to the original conversation record and possibly to each other.

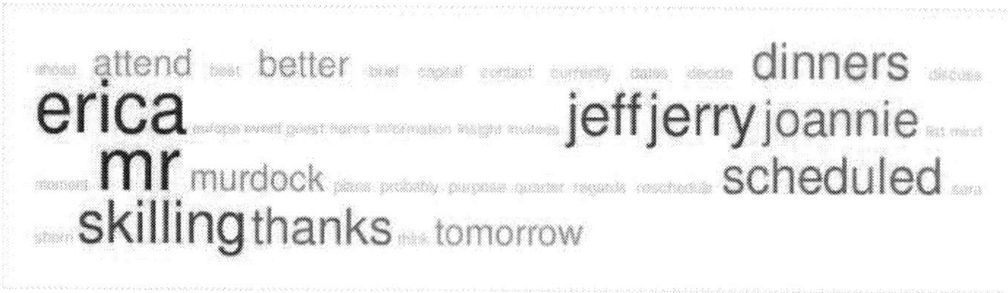

Figure 4.1: Word cloud representing email discussion.

4.3 SUMMARIZING CONVERSATIONS IN ONE DOMAIN

In the following subsections, we consider each conversation domain in turn and describe work in that area. For each domain, we first introduce and briefly describe case studies of summarization systems that have been developed. We then compare and contrast those case studies and use them as a jumping-off point for a more general discussion of critical issues in that domain.

4.3.1 SUMMARIZING EMAILS

In this section we first introduce existing work on email summarization, highlighting individual systems and techniques that have been successful and/or influential. We subsequently use those case studies to further a discussion on inputs and assumptions, measures of informativeness, and outputs and interfaces for email summarization, comparing and contrasting the systems as we go. The focus of this section will be almost exclusively on extractive techniques, as the vast majority of email summarization research has been extractive.

Email Summarization Case Studies Work on email summarization can be divided into summarization of individual emails and summarization of email threads. Muresan et al. [2001] take the approach of summarizing individual email messages, first using linguistic techniques to extract noun phrases and then employing machine learning methods to label the extracted noun phrases as salient or not. Summarization of individual emails is a useful task for email triage and for displaying incoming emails on small handheld devices, to give two examples. Since we are interested in conversational data, we will focus here on describing techniques for summarization of entire email threads.

summarizing threads

Lam et al. [2002] take an approach to email summarization that is a hybrid between single email summarization and thread summarization. Their system summarizes individual emails but in a thread-aware manner, so that the summarized email is presented with some context from the preceding email messages. Messages subsequent to the one being summarized are ignored. The system also extracts *features* from the emails, such as dates, people's names and company names, presented as a list along with the summary text. The summarization component itself is treated as a black box, with the authors testing several standard summarizers and finding little performance difference. This stands in contrast with many approaches described below, where researchers explore email-specific summarization techniques. Lam et al. conducted a small user study to gauge the perceived suitability of their summaries for several tasks: email triage, cleanup and calendaring. An interesting finding is that all of the user study participants stated that they would have liked action-oriented email summaries, indicating whether or not the email recipient needed to take some course of action.

Rambow et al. [2004] created a sentence extraction approach for email thread summarization using supervised machine learning. They employ three classes of features: *basic* features common to any text, such as sentence length and an average tf.idf score, *message* features that take into account that an email thread is divided into multiple messages, such as the position of the message in the thread, and *email features* that capture email-specific information, such as a sentence's subject-line overlap and the number of recipients for an email. Their general finding is that the supplementing the basic features with email features yields the best overall classification results.

The Rambow system is also interesting for the manner in which the summaries are presented. The extracted sentences are processed by a module that wraps each sentence in additional text, conveying information about the sender, the date and the speech act of the sentence. For example, the following extracted sentence 1 would be converted to sentence 2:

1. Are you sending out upcoming events for this week?
2. In another subthread, on April 12, 2001, Kevin Danquoit wrote: Are you sending out upcoming events for this week?

This wrapper module has the potential to increase the coherence of the extractive summary, which otherwise could suffer from the fact that its concatenated sentences have been removed from their original contexts. However, the authors did not evaluate the impact of this wrapper text. The wrapper module can be seen as a nod towards abstractive summarization, since there is new text describing the email content at a higher level. More precisely, this is a form of hybrid extractive/abstractive summarization .

coherence

hybrid summarization

Whereas the Rambow et al. system is supervised, Newman and Blitzer [2003] present an *unsupervised* approach for summarizing very long email newsgroup conversations. The approach rests on first clustering discussion messages by topic and then extracting sentences for each cluster. Initially, each message belongs to its own cluster, and at each step of the clustering process two clusters are combined if they are connected by the most similar sentence pair. Once clustering is completed, sentences are selected from each cluster based on a variety of scores, which include the use of email-specific features pertaining to the thread structure and quoted text. For example, a sentence from a particular email message is more likely to be considered important if it is subsequently quoted in other messages. This exploitation of quoted text is a classic example of *email-specific* summarization, with the intuition being that sentences that are quoted in subsequent emails are likely to be important.

quoted text

While Newman and Blitzer focus on newsgroup discussions, Wan and McKeown [2004] focus on summarizing another particular type of email discussion, where the conversation represents a decision-making process. The system works by identifying an issue sentence in the originating email and extracting reply sentences to that issue from subsequent emails, for each participant. The issue sentence is determined by comparing each candidate sentence vector in the originating email to a comparison vector representing all replies, with the issue sentence being most similar to the comparison vector. The authors consider several ways of constructing the comparison vector, such as a standard centroid (a feature vector representing the document, with significant terms represented using term-weights such as *tf.idf* or normalized frequency), a centroid with singular value decomposition applied (mapping the sentences to a lower dimensionality, revealing core "concepts") and a combined voting approach. Response sentences are simply selected by taking the first sentence of the replies from each participant. An example summary from Wan & McKeown is shown below:

> *Issue*: Let me know if you agree or disagree w/choice of plaque and (especially) wording.
> *Response 1*: I like the plaque, and aside for exchanging Dana's name for "Sally Slater" and ACM for "Ladies Auxiliary," the wording is nice.
> *Response 2*: I prefer Christy's wording to the plaque original.

Nenkova and Bagga [2003] aim to create indicative summaries of an email thread, providing enough information about a thread to allow a user to decide whether or not to retrieve the entire thread for browsing. Each summary begins with the subject line of the root email. A sentence from the root email is then selected based on the overlap of its content terms with the subject line. The remaining summary sentences are chosen by selecting, for each reply email, the sentence that has the highest overlap of content terms with the entirety of the root email. The aim is that the resultant summary will describe the subject of the thread, a statement of a problem or information request, and a brief digest of the immediate responses to that statement. The authors found that this worked well on their particular dataset, the Pine-Info mailing list[1], because threads typically begin with a user asking for help on a particular problem and receiving numerous suggestions.

clue words

Carenini et al. [2007] created an unsupervised email thread summarization system based on *clue words*. Their approach relies on the conversation structure of the emails and the repeated words throughout the thread. The email conversation is represented as a graph structure with email fragments as nodes (i.e., the Fragment Quotation Graph introduced in Section 3.4.5). Clue words are the highly informative words that occur in adjacent nodes of the graph. This system exemplifies the general idea of representing conversations as graphs, where nodes can represent sentences, fragments or conversation participants.

Assumptions and Inputs for Email Summarization Systems As noted previously, email summarization systems differ in whether the input document to the summarizer is a single email or an email thread. Here, we consider systems that summarize partial or entire threads. Systems also differ in whether they expect a user-supplied query or solely the document to be summarized.

Many of the systems we mentioned are designed based on assumptions about the nature and purpose of the email conversations. For example, the system of Wan and McKeown [2004] assumes

[1] `http://www.washington.edu/pine/pine-info/`

that the input conversation will be a decision-making process between multiple participants, where feedback has been solicited by the original email sender, and it is assumed that the original email contains a concise one-sentence statement of the issue at hand. The motivation for this style of email summarization is to aid the decision-making process by concisely displaying the issue and reporting the responses so far, thereby allowing a participant just joining the conversation to easily gauge the state-of-affairs and make a contribution relevant to the other responses.

Similarly, the system of Nenkova and Bagga [2003] assumes that threads begin with a user asking for help and subsequently contain multiple responses to that original request. This is a good characterization of many discussion and help forums but may not be true of the majority of email conversations. These assumptions are not weaknesses of the particular systems, but rather *specializations*.

Other systems make assumptions about the structure and formatting of the email data. For example, the system of Rambow et al. [2004] assumes that the emails contain headers and that the headers can be parsed to get information such as the subject, timestamp and the number of recipients. The system of Carenini et al. [2007] assumes that the email threads contain quoted material so that a given email might contain fragments of previous emails. The challenges of dealing with quoted text include email authors varying as to whether they use selective quotation, inline quotation, or no quotation, and the fact that quoted text is not signaled uniformly across email software programs.

Measuring Informativeness in Email Summarization Systems The measure of informativeness for a given summarization system cannot be separated from the assumptions and inputs for that system. For example, the systems of Wan and McKeown [2004] and Nenkova and Bagga [2003] give special status to the root email and one or more of its sentences, and rank sentences from the subsequent emails according to their similarity with the root, based on the assumptions described in the previous section that the root contains either an information need or a decision statement.

Supervised approaches such as the system of Rambow et al. [2004] make fewer assumptions about which structural or lexical features signal informativeness, but rather learn informativeness by training on labeled data. The informativeness of a test sentence will ultimately be gauged by a combination of its lexical, structural and email-specific features and how these features correspond with the learned model. In this supervised framework, the measure of informativeness depends on what it is the annotators who created the gold-standard were labeling. If they were labeling generically informative sentences, the system will generate generically informative summaries. However, one could generate decision-oriented or action-item oriented summaries by training on data labeled with decision and action-item sentences. Similarly, one could create a "sentimental summary" by training on sentences labeled for positive- and negative-subjective sentences. In other words, supervised systems are still built on assumptions about which phenomena are interesting and which should therefore be learned.

Many systems, e.g., Newman and Blitzer [2003], measure informativeness by first identifying topics and then selecting sentences that are representative of each topic. These systems assume that an email thread contains multiple topics and that good candidate sentences exist for describing each topic. Techniques for topic identification and segmentation are described in Chapter 3.

This discussion highlights the interplay between the *assumptions and inputs* for a system and the system's *measures of informativeness*. Each of the mining techniques described in Chapter 3 can be considered a potential input to a summarizer, and the summarizer can use one or more of these inputs to measure informativeness. For example, a user might desire an email thread summary that lists action items according to topic, requiring a system that uses action-item and topic detection results and combines them in a meaningful way.

subject-line overlap

A common feature used for measuring informativeness in email summarization is subject-line overlap or similarity (e.g., Rambow et al. [2004] and Nenkova and Bagga [2003]). For this approach to be effective, one assumes that: (a) the thread contains an informative subject line; (b) the thread topics do not stray far from the original subject line; and (c) informative sentences will be similar to the subject line either through direct lexical overlap or through a similarity measure that may include term expansion. Comparing sentence similarity or overlap with the subject-line is similar to techniques in *query-dependent summarization* where one evaluates not only how informative a sentence is on its own but how similar it is to a user-provided query. Figure 4.2 shows an example of email sentences overlapping with the subject line.

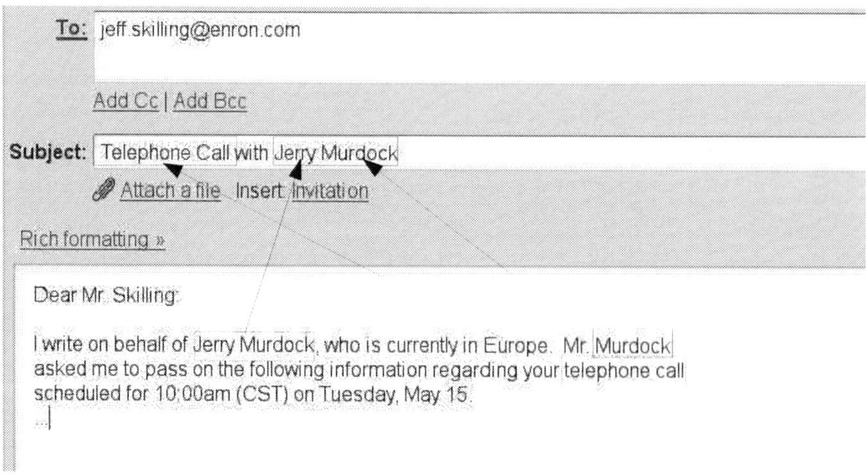

Figure 4.2: Using subject-line overlap for measuring informativeness.

Outputs and Interfaces for Email Summarization Systems All of the email summarization systems described in the case studies are extractive, meaning they ultimately output a list of sentences from the thread. Indeed, the vast majority of work on email summarization has been extractive. The system of Rambow et al. [2004] can be seen as a step towards abstraction, as it generates wrapper text for the extracted sentences. This wrapper text is intended to provide context and increase coherence for the otherwise cut-and-paste summary. The work of Lam et al. [2002] is also somewhat abstractive in that it supplements the extractive summary with a list of people's names, company names and dates.

4.3.2 SUMMARIZING MEETINGS

In this section we first describe existing work on summarizing meetings, and subsequently use those individual case studies to further the discussion of inputs and assumptions, measures of informativeness, and outputs and interfaces for meeting summarization.

Meeting Summarization Case Studies The earliest work on meeting summarization consisted primarily of applying text summarization techniques to speech transcripts . Waibel et al. [1998] implement a modified version of MMR (defined in Section 4.2.2) for summarizing meetings, presenting the user with the *n* best sentences in a meeting browser interface. The browser contains several information streams for efficient meeting access, such as topic-tracking, speaker activity, audio/video recordings and automatically-generated summaries.

speech transcripts

In more recent work, researchers began investigating how to supplement lexical information with features derived for the speech signal and features characterizing meeting structure. Murray et al. [2005a] compare unsupervised text summarization approaches such as MMR with supervised approaches incorporating prosodic features such as pitch and energy, with human judges favoring the feature-based approaches. They report ROUGE-1 recall scores in the range of 0.55-0.69 when comparing automatic extracts with gold standard *abstracts*. However, they also show [Murray et al., 2005b] that ROUGE scores did not correlate well with human judgments on the ICSI test corpus. In subsequent work [Murray et al., 2006], they began to look at additional speech-specific characteristics such as speaker and discourse features.

prosodic features

In a similar vein, Galley [2006] uses skip-chain Conditional Random Fields[2] to model pragmatic dependencies such as question-answer between paired meeting utterances, and uses a combination of lexical, prosodic, structural and discourse features to rank utterances by importance. The types of features used are classified as *lexical features*, *information retrieval features*, *acoustic features*, *structural and durational features* and *discourse features*. Galley finds that while the most useful single feature class was *lexical* features, a combination of acoustic, durational and structural features exhibited comparable performance according to Pyramid evaluation. Galley reports ROUGE-2 recall scores in the range of 0.42-0.44 and ROUGE-1 recall scores of 0.91 when evaluating automatic extracts in comparison with gold standard *extracts* (not gold standard abstracts), and Pyramid scores in the range of 0.504–0.554.

pragmatic dependencies

Also using the ICSI corpus, Liu et al. [2007] report the results of a pilot study on the effect of disfluencies (see Chapter 2) on automatic speech summarization. They find that the manual

[2]Conditional Random Fields (CRF) are undirected graphical model which have been successfully applied to label a sequence (chain) of observations, for instance a sequence of sentences in a document, with appropriate labels, for instance informative vs. non-informative sentences. In a standard CRF, the label of an observation can only influence the label of the next observation in the chain, in Skip-Chain Conditional Random Fields the label of an observation can also influence labels further down the chain (by "skipping" the intermediate labels). See Sutton and McCallum [2004] for more details.

removal of disfluencies did not improve summarization performance according to the ROUGE metric. In related disfluency work in another domain of conversational speech, Zhu [2006] show how disfluencies can actually be exploited for summarization purposes and find that non-lexicalized filled-pauses were particularly effective for summarizing SWITCHBOARD conversations. ROUGE-1 scores range between 0.502 for 30% utterance-based compression to 0.628 for 10% compression.

Liu et al. [2010] have more recently researched the impact of piping ASR output to a text summarization algorithm such as MMR, but varying n in the n-best hypotheses output from the speech recognizer to the summarizer. A typical ASR transcript is the 1-best hypothesis of the recognizer, i.e., the single "best guess," but one can also use less probable hypotheses. They found that ROUGE-2 F-scores improve from about 0.25–0.27 when using a value of n greater than 1.

Much work has been done on meeting speech which is not called automatic summarization, but which nonetheless can be considered a kind of focused summarization. For example, work on decision detection and action item detection aims to identify and extract sentences that contain or relate to decisions and action items, respectively. Using the output of a decision classifier or action item classifier, one could easily generate an extractive summary focused only on a particular phenomenon of interest. Chapter 3 describes decision and action item detection in detail.

Kleinbauer et al. [2007] present an abstractive summarizer for meetings. This system utilizes automatic topic segmentation and topic labels, and finds the most commonly mentioned content items in each topic. A sentence is generated for each meeting topic indicating what was discussed, and these sentences are linked to the actual dialogue acts in the discussion. These summaries rely on *manual* transcripts. The summarizer is not fully automatic, as it also relies on manual annotation of propositional content. However, this can be considered one of the first projects focused on moving beyond sentence extraction for summarizing meetings.

Assumptions and Inputs for Meeting Summarization Systems Most of the meeting summarization systems described above assume that an ASR transcript is available. While the AMI and ICSI meeting corpora contain both manual and ASR transcripts, and researchers experiment with both types in order to make comparisons, deploying a meeting summarization system in the real world would entail an ASR component. This is not necessarily a system development bottle-neck, as off-the-shelf speech recognizers are available, but the high word-error rates (WER) of ASR systems applied to multi-party speech do pose a significant problem. For example, the AMI ASR system [Hain et al., 2007] features a WER of approximately 38%. This high WER is less problematic for the effectiveness of the systems themselves – it is a consistent finding that summarization performance does not greatly degrade on ASR transcripts [Murray et al., 2005a] – than for the end user who may have difficulty reading summaries of noisy transcripts.

Some meeting summarization systems might assume that there is further instrumentation of the meeting room, e.g., video recordings, slide capture, whiteboard events and electronic notes. These types of multi-modal data are available as part of the AMI corpus, but such complex instrumentation may not always be feasible in a real life scenario.

Systems such as that of Kleinbauer et al. [2007] assume that the meetings follow a particular scenario, with the participants having distinct roles and the group working together towards a specific goal. Each meeting in the scenario represents a particular design stage. The summarizer can create rich and detailed abstracts for meetings that follow such a scenario. However, applying the system to other types of meetings and conversations would require significant effort in terms of ontology design and retraining, etc.

meeting scenarios

Measuring Informativeness in Meeting Summarization Systems As evidenced by the case studies, meeting summarization systems have typically taken one of two general approaches: feeding an ASR transcript to a text summarization algorithm such as MMR, or using more speech-specific approaches that may incorporate prosody and dialogue features. Penn and Zhu [2008] question the true impact of "avant-garde" features such as speech prosody, showing that much of the improvement those features brought could be captured by much simpler features capturing the length or duration of each utterance. Similarly, Murray [2007] separates length and duration features from "true" prosodic features and finds that length features are indeed a challenging baseline. However, it is also found that one can achieve respectable extractive summarization results, with AUROC scores as high as 0.74, using *only* true prosodic features such as energy and pitch and no use of lexical or structural features.

avant-garde features

In our later discussion on summarizing conversations across modalities in Section 4.4, we will again see that—similar to the findings of Zhu and Penn—a competitive system need not incorporate domain-specific features such as prosody. But in situations where a transcript might not be available, it is interesting that prosody alone can be useful for indicating informativeness, and one could generate an audio summary using only features from the speech signal.

Beyond prosody and dialogue features, there has been little work on investigating the use of other "avant-garde" features available from the multi-modal datastream, such as notes, slides, and whiteboard events. It remains to be seen how big of an impact these features might have on summarization performance.

Outputs and Interfaces for Meeting Summarization Systems With meeting summarization, there is a great number of possible outputs and interfaces. While informativeness might be determined as discussed in the previous section, using perhaps a variety of text and speech features, the summary output could be completely non-textual in order to minimize the exposure of end-users to noisy ASR data. For instance, the summary could be a concatenation of the relevant audio clips, or a video summary .

video summary

Otherwise, with meeting summarization, extractive systems are at a potential disadvantage compared with abstractive systems, as the summary units will be disfluent utterances taken from the noisy, error-filled ASR transcript. Even if the sentence classification is good, readers may find it very tedious or difficult to read the extractive summary. A simple way to improve a meeting extract is to remove filled pauses and try to repair some disfluencies.

disfluency removal

meeting browsers

In much of the work on meeting summarization, the summaries are meant to be a component of a meeting browser, serving as an index into the audio-video meeting record. The summaries may be time-aligned with other artefacts from a meeting such as notes, slides and visualizations of speaker activity. In these cases, extractive summary sentences will almost always have a one-to-one mapping with transcript sentences, while abstract sentences can have a many-to-many mapping with the transcript. This highlights the fact that abstractive systems are identifying patterns, messages or events that aggregate numerous sentences. Figure 4.3 shows an example of a meeting browser that incorporates abstract summaries, with the summary linked to the transcript and the transcript time-aligned with the audio and video.

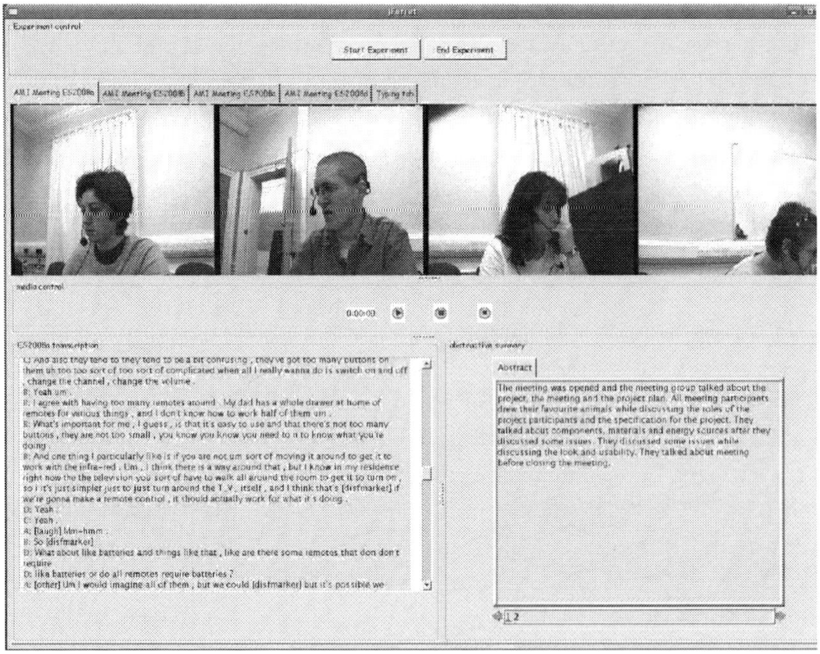

Figure 4.3: Meeting browser incorporating abstractive summaries.

One alternative to summarizing meeting speech is simply to speed it up. Experiments have shown users still have good comprehension of meeting discussions even when the discussion is played several times faster than the original speed [Tucker and Whittaker, 2006]. However, user satisfaction is not high.

4.3.3 SUMMARIZING CHATS AND BLOGS

In this section we first present case studies of summarization applied to online chats and blogs, and then discuss assumptions and inputs, measures of informativeness, along with outputs and interfaces in these contexts.

Chat and Blog Summarization Case Studies In Zhou and Hovy [2005], the authors address the task of automatically summarizing internet relay chats, using online discussions pertaining to the GNU/Linux project[3]. These discussions actually consist of both chats and emails. An interesting facet of the data is that the online community provided digests of its own discussions, including quotes and hyperlinks. These served as naturally occurring gold standard summaries for training and evaluation purposes.

The approach presented in Zhou and Hovy [2005] is to first segment and cluster the message data, and then identify *adjacency pairs* in the text (see Chapter 3 for more details on these tasks). An example of an adjacency pair is a question-answer pair, where a person raised a question and another person subsequently answered that question. A mini-summary is generated for each topic, where a topic is represented by a cluster of messages. Each mini-summary consists of an initializing segment of an adjacency pair followed by one or more responding segments. A supervised approach is taken, comparing maximum entropy and SVM models [Poole and Mackworth, 2010] with simple lexical and structural features. The SVM classifier was found to outperform the maximum entropy model.

adjacency pairs

Hu et al. [2007] is one of the first examples of blog summarization to consider the blog comments as an informative piece of data. While the extractive summaries they generate are summaries of the original blog posts only, the authors weight blog post sentences highly if their constituent words appear often in widely quoted comments and are often used by *authoritative* readers. They also conducted a user study and found the interesting result that human summary annotators will change their sentence selection decisions for a blog post when they are allowed to read blog comments in addition to reading the blog post itself. This tells us that blog commenters play a large role in highlighting, and even determining, what is most salient in a blog post.

blog comments

Whereas Hu et al. consider longer blog posts and comments, in the work of Sharifi et al. [2010], the task is automatically summarizing microblogs such as Twitter messages. Given a trending (i.e., currently popular) topic and a set of messages, or *tweets*, concerning that topic, their system produces a very brief, one-sentence summary of the topic. This data is not conversational in the sense of meetings, emails or chats, where participants are directly engaging and responding to one another, but it is conversational in that many thousands of users are simultaneously tweeting about the same topic, forming a massive community conversation. The purpose of the summary is to concisely convey why the topic is trending. For example, the topic *Ted Kennedy* might be trending because Ted Kennedy died, and so the system will generate a summary such as *A tragedy: Ted Kennedy died today.* The algorithm takes as input a topic phrase (e.g., *Ted Kennedy*) and a set of sentences from relevant tweets, and builds a graph representing word sequences that occur before and after the topic phrase. Individual word nodes are weighted according to their occurrence count in that position and their distance from the root node. Generating a summary consists of finding the path with the largest weight from the root topic phrase to a non-root node. The root node is reinitialized with this partial path and the rest of the sentence is generated by again finding the path with the largest weight from this new root node to a non-root node. One effect of this implementation is

[3]`http://www.gnu.org`

that the generated summary sentence will always be a string that actually occurred in at least one of the input sentences; in other words, it is an extractive summary.

In 2008, the Text Analysis Conference[4] ran a pilot task on summarizing opinions in blog posts. However, the blog data was not conversational in nature but rather featured individual blog posts on a set of topics. For that reason, we do not describe the task or the submitted systems in detail here since our primary interest is conversations. It could be argued, however, that the set of posts on a given topic are conversational in a very general sense, since they feature people blogging on a common topic and possibly reading each other posts, albeit not replying to each other in an interactive fashion; this is "conversational" in the same sense that the multitude of Twitter posts on a given topic in Sharifi et al. [2010] is conversational.

Assumptions and Inputs for Chat and Blog Summarization Systems Summarization of online conversations such as blogs and forum discussions has been a less researched area than meeting and email summarization, for a variety of reasons. Firstly, online conversations have only become widely popular in recent years. Secondly, a point related to the first, is that there is no large, publicly available blog corpus annotated with extractive and abstractive summaries such as exist for meetings (the AMI and ICSI corpora) and emails (the Enron and BC3 corpora). And thirdly, the inputs, tasks and use cases are not clearly defined nor agreed upon.

In the work of Zhou and Hovy [2005], the data are very conversational and personal, with participants responding directly to one another. This contrasts with the work of Sharifi et al. [2010], where the conversation is very diffuse and spread out; people are discussing a common topic on a massive scale, sometimes responding directly to one another but often not. Somewhere in between is the work of Hu et al. [2007], where the goal is to summarize individual blog posts in the context of the comment discussions. These are all large-scale, online conversations, but the datasets are structured differently from one another and the summarization goals are different as well.

It would be advantageous for the research community to define clear blog summarization tasks, and to facilitate the creation of an annotated blog summarization corpus related to the tasks of interest. A well-defined task could involve summarizing blog posts themselves, blog comments, blog links, or some combination thereof. Blogs are an interesting case because there are typically several types of conversations happening. In a group blog, the bloggers may be posting in response to one another. In their individual posts, commenters may be carrying on discussions. Both bloggers and commenters may also be linking to outside sources, forming a wider conversation.

Measuring Informativeness for Chat and Blog Summarization Systems Informativeness in both Zhou and Hovy [2005] and Hu et al. [2007] is measured by considering not just the initiating post, but responses to the post as well. In the case of Zhou and Hovy [2005] this is done by identifying adjacency pairs in chat conversations, while in Hu et al. [2007] the researchers use blog comments to consider the informativeness of the original post sentences. While the initial post may

[4] `http://www.nist.gov/tac/2008/`

contain the bulk of the discussion, in each case it is the ensuing conversation that illuminates what is important.

In Sharifi et al. [2010], the conversation is of a massive scale, with hundreds or thousands of participants talking about a particular person, event or thing. Informativeness is determined by analyzing the many lexical patterns used to refer to that topic. The final summary is a crowd-sourced digest of the trending item.

A source of information available for online conversations is the presence of links within the conversation. Sentences could be weighted not only by their lexical and structural features within the given conversation, but also by whether they link to other documents and by the content of those other documents.

Outputs and Interfaces for Chat and Blog Summarization Systems If the input conversation has a threaded structure, an extractive summary of the conversation will likely need to be threaded as well to maintain coherence. An abstractive system might generate separate paragraphs for each thread, or try to aggregate similar threads.

Once an online conversation exceeds a certain size in terms of participants and number of comments, extraction alone is probably not feasible to accurately characterize the discussion. One strategy is to aggregate similar comments and generate new text to describe them, while presenting a few of the original sentences as examples. This would constitute a hybrid extractive-abstractive system.

For massively large online conversations, visualizations can also be a good complement to the text. For summarizing thousands, or even millions, of tweets, a mix of information visualizations and word clouds, such as Figure 3.7 can be very effective. One could also visualize clusters of conversation participants to easily see which individuals are interacting the most.

4.4 SUMMARIZING MULTI-DOMAIN CONVERSATIONS

The summarization systems discussed up until this point have primarily been designed with particular domains in mind and attempt to harness unique features of those domains. For example, meeting summarization systems often use prosodic features while email summarizers derive metadata from the email headers. In contrast, other summarization systems have been designed to work across a variety of conversation domains and modalities. Here, we briefly discuss several such multi-domain systems.

In early work on conversation summarization, Zechner [2002] investigates summarizing several genres of speech, including spontaneous meeting speech. While this work focuses on spoken modalities, the system is not speech-specific and could be applied to written conversations as well. Though relevance detection in his work relies largely on *tf.idf* scores, Zechner also explored cross-speaker information linking and question-answer detection, so that utterances can be extracted not only according to high *tf.idf* scores, but also if they were linked to other informative utterances. This work also focuses on detecting disfluencies such as filled pauses, false starts and repairs in order to increase summary readability and informativeness.

Murray and Carenini [2008] have developed extractive summarization techniques for spoken and written conversations, with a focus on meetings and email. The primary goal in this research is to identify a common feature set that would yield good summarization performance in both spoken and written conversations. The features included speaker/participant dominance, lexical cohesion, participant-based term weights, centroid similarity scores, turn-taking and other structural characteristics of multi-party conversation. The participant-based term weights are based on the intuition that certain words will tend to associate more with some participants than others, owing to varying roles (e.g., industrial designer vs. financial expert) and generally different areas of interest and expertise. When rating the features according to the F statistic (basically the ability of an individual feature to discriminate the positive and negative classes) they find that the feature rankings are very similar for the two domains. This is shown in Figures 4.4 and 4.5 (source: [Murray and Carenini, 2008]). We do not discuss the individual features in detail here, but merely note that the most important feature subsets are very similar in the two domains.

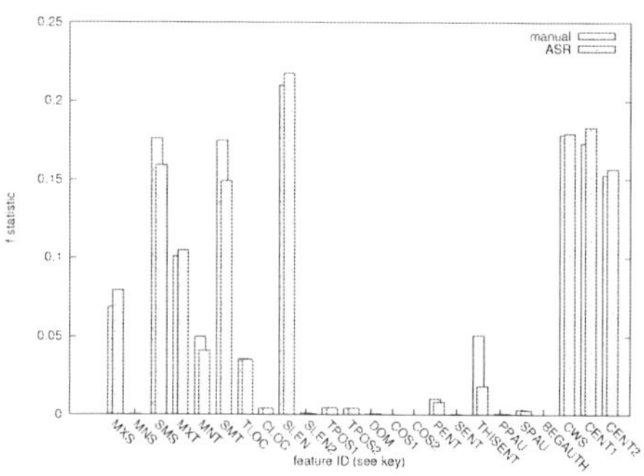

Figure 4.4: F statistics, meetings.

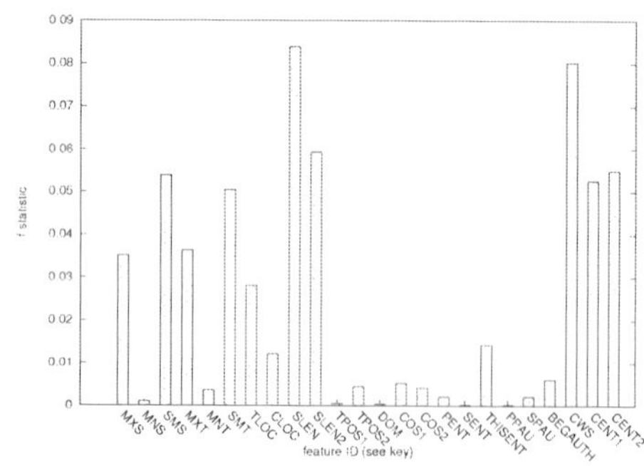

Figure 4.5: F statistics, emails.

The highest AUROC scores for the extractive classifiers are approximately 0.85 for meetings and 0.75 for emails. Most significantly, they compare the conversation-features approach to a speech-specific system for meetings and an email-specific system for emails and find that the conversation-features approach performs just as well in each domain. That is, there was no bonus to using domain-specific features and one could instead rely on general conversation features. This finding is similar to those of Penn and Zhu [2008], who find that "avant-garde," domain-specific features often provide little or no performance improvement over more general features (see Section 4.3.2).

domain adaptation

Sandu et al. [2010] also use general conversation features and try to leverage the large amount of available labeled meeting data to improve summarization results in the less-resourced domain of emails. This is the general problem of *domain adaptation*, where one tries to adapt a system developed in a "source" domain to data in a "target" domain. The authors found that domain adaptation techniques were helpful when no labeled email data was available.

4.4.1 ABSTRACTIVE CONVERSATION SUMMARIZATION: A DETAILED CASE STUDY

Recent work on summarizing multi-domain conversations has taken a more *abstractive* approach, generating novel text to describe the conversation rather than extracting sentences from the conversation itself. We will describe one system in considerable detail to see how abstractive systems differ from the extractive systems described previously. In the system of Murray et al. [2010], the abstractive summarizer proceeds in a pipeline of *interpretation*, *transformation*, and *generation*. We can first describe each of these stages at a high level:

- *Interpretation.* Mapping the input conversation to a source representation.
- *Transformation.* Transforming the source representation to a summary representation.
- *Generation.* Generating a summary text from the summary representation.

interpretation

At a system level, the Murray et al. abstractive system carries out interpretation by mapping conversation sentences to a simple conversation ontology written in OWL/RDF. This entails populating the ontology with instance data corresponding to the particular conversation participants, the entities or topics discussed, and dialogue-acts such as decisions being made, problems encountered, and opinions expressed. These latter sentence-level phenomena are determined using supervised classifiers and a variety of structural, lexical and conversation features. The interpretation stage also involves detecting *messages*, which are essentially collections of sentences which mention the same entity, belong to the same participant and have the same dialogue act type. That is, a message is an *abstraction* over multiple sentences.

transformation

The transformation stage is responsible for selecting the most informative messages. The content selection is carried out using Integer Linear Programming (ILP), where a function involving message weights and sentence weights is maximized given a summary length constraint. Messages are weighted according to the number of sentences they contain (i.e., roughly how much information they express), while sentences are weighted according to their posterior probabilities derived from the supervised classifiers in the preceding interpretation stage (i.e., the predictions of decisions, actions, problems and sentiment). The idea is that sentences relating to these types of phenomena should be included in the summary. The output of the transformation stage is simply a set of messages.

generation

The generation stage takes those selected messages and creates a textual summary by associating elements of the ontology with linguistic annotations. For example, participants are associated with an identifier such as their name, email or role in an organization. Topics or entities are simply weighted noun phrases from the conversation. An individual summary sentence is realized by associating a verbal template with the message type. For example, instances of DecisionMessage are associated with the verb *make*, have a subject template set to the noun phrase of the message source (the participant), and have an object template *[NP a decision PP [concerning]]* where the object of the prepositional phrase is the noun phrase associated with the message target.

This system architecture is very similar to data-to-text systems such as described in Portet et al. [2009] and more generally in Reiter and Dale [2000], with the primary difference being textual input rather than raw data such as numeric measurements (in their case, turbine readings, weather stations, and intensive-care unit monitors). In either case, one is looking for messages (or patterns) in the input, selecting the most critical messages, structuring and combining them in a coherent fashion, and finally generating text to describe the body of selected messages.

To give a concrete example of how an abstractive conversation summarizer such as that of Murray et al. [2010] compares with an extractive system such as Murray and Carenini [2008], we can again consider the sample conversation presented at the beginning of this chapter. For the purpose of this example, let us assume that a three-sentence summary of the conversation is required. Based on features such as sentence position, sentence length, and term-weights, the following three sentences could be selected by the extractive system:

- **From Erica**: The purpose of your conversation with Mr. Murdock is to discuss the above in more detail and to more fully brief you on the purpose of these dinners.
- **From Joannie**: Would it be possible to schedule during the next quarter?
- **From Erica**: Given the time difference involved I will not be able to contact him before the appointed time tomorrow, and therefore I'd very much appreciate if we could go ahead with the call as planned.

In contrast, the abstractive system would first populate the ontology with participant instances (Joannie, Erica, Sherri) and entities (e.g., phone call, office, vacation). Sentences would then be analyzed using dialogue act classifiers. For example, the following sentence is determined to be a subjective sentence:

Joannie: Erica, Due to the fact that Jeff is unable to attend on July 19, I believe it would be better to reschedule the call for sometime next quarter.

The abstractor would then look for patterns of similar sentences and combine similar sentences into a message. For example, sentences where Joannie is making negative-subjective comments about the phone call could be combined into a single message with *Joannie* as the message source and *phone call* as the message target. A subset of messages would be selected for the final summary. The summary text could look something like the following.

Erica mentioned an action item concerning the phone call. Joannie then expressed some negative opinions about the scheduling. Finally, Sherri asked a question about the upcoming dinners.

In both summary cases, the summary sentences can be linked to the original document sentences. This allows the reader a better understanding of the context and the surrounding sentences. Moreover, the summary essentially serves as a gateway for the reader to more systematically browse through the original sentences. For instance, the reader may be very interested in browsing through all the sentences expressing negative opinions, sentences that represent action items, or sentences that describe decisions made. While the underlying conversational data are unstructured, the summary sentences and the linking to original sentence essentially provide structured metadata to access the underlying data.

A potential weakness of extractive summaries is that they can lose coherence since the summary sentences are removed from their original contexts. For example, a person reading the short extract above may not know who Mr. Murdock is since the original preceding sentence is missing. On the other hand, an abstractive summary can be too general. For example, a person reading the short abstract above might want to know what Joannie's negative opinions were specifically. In each case, we can supplement the summaries by linking sentences back to the conversation.

Murray et al. [2010] have carried out a user study comparing extractive and abstractive summaries of meeting conversations. Participants rated automatically generated abstracts significantly higher than even *human*-selected extracts on several usability and readability criteria. In qualitative comments, many participants stated that they did not even consider the extracts to be summaries and were therefore "useless." However, some participants criticized the automatically generated abstracts for being too vague and linguistically repetitive.

4.5 CONCLUSION

In this chapter, we discussed summarization of conversations in domains such as meetings, emails and blogs. We have discussed domain-specific approaches as well as summarization approaches that can be applied to conversations in diverse modalities. In presenting our case studies, we have presented an analysis in terms of *inputs and assumptions*, *measures of informativeness* and *outputs and interfaces*.

Findings by Penn and Zhu [2008] and Murray and Carenini [2008] suggest that creating domain-specific conversation summarization systems is not always worthwhile. A suitable approach is to model conversations more generally and exploit features common to all multi-party interactions. Such systems will also have the benefit of being easily extendible to novel conversation modalities that will come about due to technological change.

According to intrinsic evaluation measures, the performance of extractive summarization systems is nearing human-level performance in some domains. However, studies suggest that users are not always fond of actually reading and using extractive summaries even when they are, in fact, human-selected sentences. We believe that findings such as the user study of Murray et al. [2010] provide a great deal of motivation for doing further work on abstractive summarization, even if early attempts show that such systems are still in need of refinement.

Putting these findings together, we conclude that a promising vein of research for conversation summarization is to develop abstractive or hybrid techniques that apply to conversations across modalities.

4.6 IMPORTANT POINTS

- We discuss summarization systems according to three aspects: assumptions and inputs, measures of informativeness, and outputs and interfaces.
- The *assumptions and inputs* aspect includes the nature of the corpus to be summarized, representations of the data, and the presence of upstream processes on which summarization depends.
- The *measures of informativeness* aspect is concerned with how salience is determined within the summarization system.
- The *outputs and interfaces* aspect concerns the modality and structure of the final summary.
- We surveyed summarization work applied to meetings, emails, blogs and chats. In each of these areas, *domain-specific* summarization approaches have been developed, incorporating features and techniques specific to the particular characteristics of the data.
- *Multi-modal conversation summarization* techniques have been developed and can be applied to conversations in any modality. General findings indicate that these approaches are competitive with domain-specific techniques.
- We provided a detailed case study of one *abstractive conversation summarization* system, illustrating how such a system differs from more common extractive systems.

4.7 FURTHER READING

Several books have been published on general automatic summarization [Endres-Niggemeyer, 1998, Mani, 2001a, Mani and Maybury, 1999]. While none of these books are current, each provides an overview of the basic tasks and distinctions. Within Mani and Maybury [1999], Karen Spärck-Jones has an influential paper on factors and directions for summarization [Jones, 1999].

A more recent discussion of automatic summarization can be found in Jurafsky and Martin [2008]. This includes summarization case studies from prior to 2008.

A forthcoming book will discuss speech summarization specifically [Penn and Zhu, Forthcoming]. Whereas we concentrate on textual conversations, including spoken conversations with written transcripts, Penn and Zhu describe summarization systems that exploit speech-specific characteristics such as prosody.

CHAPTER 5

Conclusions / Final Thoughts

In the last 20 years, following the creation of the Web, a dramatic change in how people communicate has occurred. While in the past, almost all human conversations were in spoken form, nowadays more and more people are having conversations by writing in a growing variety of social media. The numbers are staggering: billions of emails are exchanged every day, there are millions of bloggers, one billion people have instant messaging accounts, and Facebook has half a billion users. In the same period of time, we have also witnessed a rapid progress in speech recognition technology, which is enabling the development of computer systems that can automatically transcribe any spoken conversations.

The net result of these two ongoing revolutions is that an ever increasing portion of human conversations is now available in *text* form, either because they were originally written, or because they were originally spoken and then automatically transcribed.

In this book, we discussed how all these text conversations can be processed by adapting Natural Language Processing (NLP) techniques originally developed for monologues, e.g., newspapers and books, or by harnessing techniques created specifically for conversational data.

More specifically, we tried to give an overview of the work currently being done on mining and summarizing text conversations, as well as highlighting promising avenues of future research. We have concentrated primarily on meetings, email and blog conversations, but rapid technological change means that we will soon be having conversations in ways that we cannot yet imagine. NLP techniques that aim to understand and summarize conversations generally are well poised to inhabit those new research spaces.

The upper-right quadrant of Figure 5.1 shows the research space that we believe is the most promising for future work on summarizing conversations, as well as being the least researched. We advocate for summarization systems that are increasingly abstractive and less restricted to particular domains. Such a vein of research will bring us closer to realizing the goal of flexible, human-style summarization of conversations spanning different domains.

The majority of summarization research today would be mapped into the upper-left and lower-left quadrants of the figure, representing extractive systems at varying levels of domain-specificity. We discussed the findings that domain-specific approaches often bring little or no benefit compared with general summarization approaches, and so, all things being equal, it would be most beneficial if any remaining work on extraction was as general as possible and focused on making such techniques increasingly abstractive, even if only at the relatively superficial level of sentence compression and aggregation. While the upper-right quadrant represents gold-standard summarization, the most realistic route to that goal may be found by filling in the lower-right quadrant. This will involve

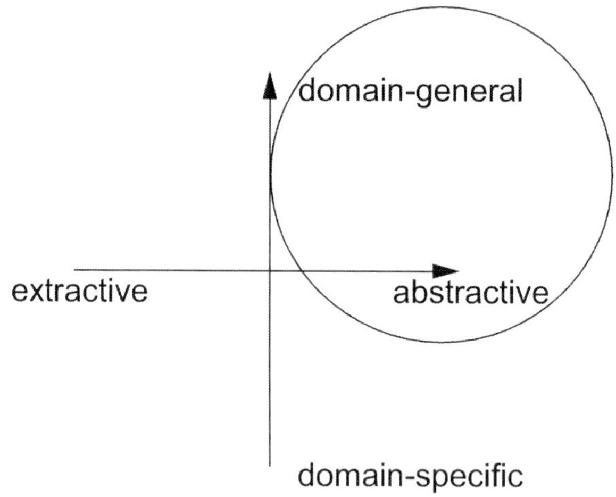

Figure 5.1: The promising research space for summarization.

developing systems that are fully abstractive, incorporating natural language generation components, but limited by domain. Such summarizers will include components able to exploit real-world knowledge about particular scenarios or applications.

If we now move from the summarization research landscape to the space of computational techniques, we envision that more powerful approaches will be developed by exploring the upper-right quadrant of Figure 5.2. In particular, when we consider the machine learning techniques on the x-axis, there is a growing interest in semi-supervised and unsupervised machine learning techniques that can be easily applied to new domains by leveraging large amounts of readily available unlabeled conversational data. Also, the need to transfer data and insights from one domain to another will generate more and more attention towards domain adaptation methods. With respect to the y-axis, the features used in the machine learning techniques, an ongoing successful trend is to expand the set from simple lexical and syntactic features to more sophisticated semantic and conversational features of the sentences (or turns). More and larger annotated corpora are also sorely needed to train and test the data-driven approaches. As we pointed out in Chapter 2 the research community would greatly benefit from the annotation of blog corpora.

Finally, when we look at the different text mining tasks, we believe that the most promising line of research is to integrate all these tasks in a mutually beneficial way. For instance, Figure 5.3 sketches a proposal in which Dialog Act Modeling, Topic Modeling and the construction of the Fragment Quotation Graph are performed simultaneously and interdependently. To what extent these tasks can be integrated and what computational framework could support such integration are key, open questions for future research.

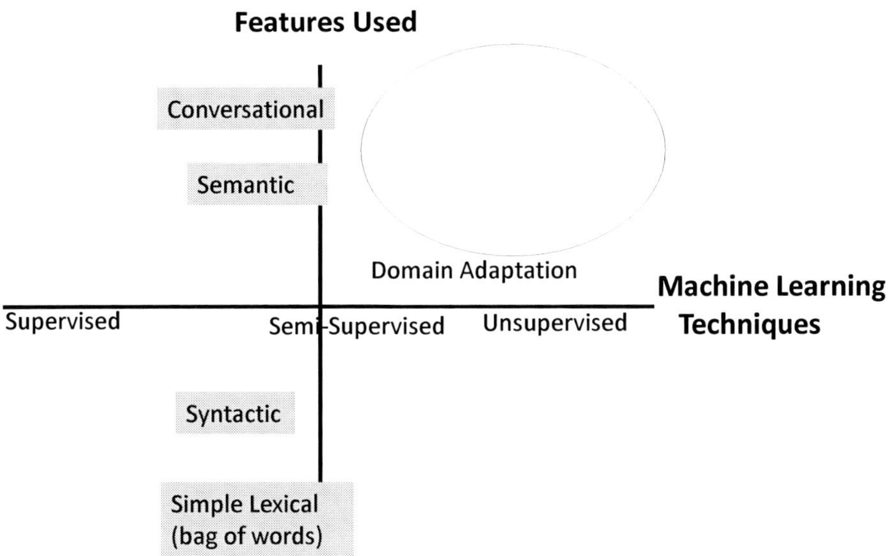

Figure 5.2: The most promising computational approaches.

To review our overall discussion of the material presented in this book, in Chapter 2 we described some of the conversation corpora that are available and widely used by researchers. The fact that there are so many freely available corpora makes this an excellent time to be researching conversation-related topics, and bodes well for future research. There is also growing agreement in this research community on the evaluation metrics and annotation standards used. For example, both the AMI and BC3 corpora contain abstractive and extractive summary annotations that are linked to one another. This mapping between extracts and abstracts can help researchers build systems that are increasingly abstractive rather than solely cut-and-paste. And having corpora in different domains that are annotated similarly to one another can aid work on domain adaptation.

In Chapter 3, we considered several text mining tasks. We started by covering tasks that can be performed on any document such as determining what topics are covered in the conversation (i.e., topic modeling), as well as detecting what opinions are expressed on those topics (i.e., sentiment and subjectivity analysis). Then, we focused on tasks that consider unique features of human conversation. In particular, we have discussed how to determine what dialog acts are expressed in the different turns, what is the thread structure of a conversation, and what turns are expressing decisions and action items. For all these tasks, we described and compared both supervised and unsupervised

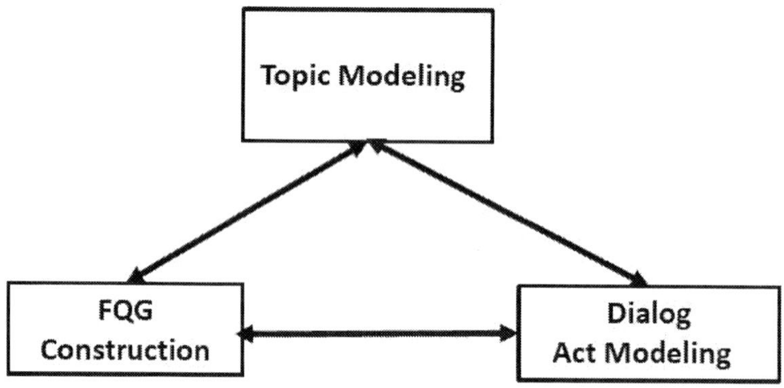

Figure 5.3: Integration of different text mining tasks.

machine learning methods. We also noticed that often synchronous vs. asynchronous conversations present different challenges. For example, synchronous conversations, especially written ones like chats, often require disentanglement to determine reply-to relationships between turns. In contrast, an interesting problem for asynchronous conversation with quotation is to extract a finer level conversational structure. Another issue we discussed is how each mining task can rely on the others; for example, work on action item detection can use predicted dialogue act labels as an input feature to a statistical classifier. Current work is exploring other beneficial dependencies among tasks; for instance, between both topic modeling and conversational structure.

In Chapter 4, we gave an overview of many existing systems for summarizing meetings, emails, blogs and forums. We examine these systems in the light of three considerations: *assumptions and inputs*, *measures of informativeness*, and *outputs and interfaces*. We hope that having the overview structured in those terms will help researchers who are considering building a summarization system in a particular domain and are uncertain of the options and requirements. We also highlight systems that have been designed to work on conversations in multiple modalities and genres. Finally, we discuss early work on abstractive systems that attempt to glean a deeper understanding of the conversation and generate new text to describe it. We use a detailed case study to show a relatively simple abstractive system can be built and how it compares with a standard extractive system.

As stated in the introduction, conversations are fundamental to the human experience, and we live in a technological age where conversations are more prevalent than ever. They span modalities and can grow to include hundreds or even thousands of people. And more often than not, conversations now exist in a lasting record that can be analyzed, mined, condensed and visualized. We hope that this book provides a glimpse of the many possibilities for summarizing and mining such conversations and gives inspiration for insightful new approaches.

Bibliography

N. Agarwal and H. Liu. *Modeling and Data Mining in Blogosphere.* Morgan & Claypool Publisher, 2009. DOI: 10.2200/S00213ED1V01Y200907DMK001 Cited on page(s) 9

D. Allemang and J. Hendler. *Semantic Web for the Working Ontologist.* Morgan Kaufmann, 2008. Cited on page(s) 85

P. Aoki, M. Szymanski, L. Plurkowski, J. Thornton, A. Woodruff, and W. Yi. Where's the "party" in "multiparty"?: analyzing the structure of small-group sociable talk. In *Proc. 20th Conf. on Computer Supported Cooperative Work*, pages 393–402, 2006. DOI: 10.1145/1180875.1180934 Cited on page(s) 75

J. L. Austin. *How to Do Things with Words.* Harvard University Press, 1962. Cited on page(s) 77

N. Baron. *Always On: Language in an Online and Mobile World.* Oxford University Press, 2008. Cited on page(s) 2, 4

R. Bekkerman, A. McCallum, and G. Huang. Automatic categorization of email into folders: Benchmark experiments on Enron and SRI corpora. Technical Report IR-418, Center of Intelligent Information Retrieval, UMass Amherst, 2004. Cited on page(s) 26

K. Bennett, A. Demiriz, and R. Maclin. Exploiting unlabeled data in ensemble methods. In *Proc. 8th ACM SIGKDD Int. Conf. on Knowledge Discovery and Data Mining*, pages 289–296, 2002. DOI: 10.1145/775047.775090 Cited on page(s) 66

P. Bennett and J. Carbonell. Detecting action-items in e-mail. In *Proc. 31st Annual Int. ACM SIGIR Conf. on Research and Development in Information Retrieval*, pages 43–50, 2005. DOI: 10.1145/1076034.1076140 Cited on page(s) 70

D. Blei and J. Lafferty. Topic models. In A. Srivastava and M. Sahami, editors, *Text Mining: Theory and Applications*, pages 71–94. Taylor and Francis, 2009. Cited on page(s) 11, 45, 49

D. Blei, A. Ng, and M. Jordan. Latent dirichlet allocation. *Journal of Machine Learning Research*, 3: 993–1022, 2003. DOI: 10.1162/jmlr.2003.3.4-5.993 Cited on page(s) 11, 45

J. Carbonell and J. Goldstein. The use of MMR, diversity-based reranking for reordering documents and producing summaries. In *Proc. 21st Annual Int. ACM SIGIR Conf. on Research and Development in Information Retrieval*, pages 335–336, 1998. DOI: 10.1145/290941.291025 Cited on page(s) 84

G. Carenini, A. Pauls, and R.T Ng. Interactive multimedia summaries of evaluative text. In *Proc. 11th Int. Conf. on Intelligent User Interfaces*, pages 124–131, 2006. DOI: 10.1145/1111449.1111480 Cited on page(s) 18

G. Carenini, R. Ng, and X. Zhou. Summarizing email conversations with clue words. In *Proc. 16th Int. World Wide Web Conf.*, pages 91–100, 2007. DOI: 10.1145/1242572.1242586 Cited on page(s) 26, 72, 73, 88, 89

G. Carenini, R. Ng, and X. Zhou. Summarizing emails with conversational cohesion and subjectivity. In *Proc. 46th Annual Meeting Assoc. for Computational Linguistics*, pages 353–361, 2008. Cited on page(s) 59, 74

J. Carletta. Assessing agreement on classification tasks: The kappa statistic. *Computational Linguistics*, 22(2):249–254, 1996. Cited on page(s) 21

J. Carletta. Unleashing the killer corpus: experiences in creating the multi-everything AMI meeting corpus. In *Proc. 5th Int. Conf. on Language Resources and Evaluation*, pages 181–190, 2006. DOI: 10.1007/s10579-007-9040-x Cited on page(s) 22, 27

V. Carvalho and W. Cohen. On the collective classification of email "speech acts". In *Proc. 31st Annual Int. ACM SIGIR Conf. on Research and Development in Information Retrieval*, pages 345–352, 2005. DOI: 10.1145/1076034.1076094 Cited on page(s) 63

A. Chapanond, M. Krishnamoorthy, and B. Yener. Graph theoretic and spectral analysis of enron email data. *Journal of Computational & Mathematical Organization Theory*, 11(3):265–281, 2005. DOI: 10.1007/s10588-005-5381-4 Cited on page(s) 26

F. Choi. Advances in domain independent linear text segmentation. In *Proc. 1st Annual Meeting North American Assoc. for Computational Linguistics*, pages 26–33, 2000. Cited on page(s) 45

W. Cohen, V. Carvalho, and T. Mitchell. Learning to classify email into "speech acts". In *Proc. 2004 Conf. Empirical Methods in Natural Language Processing*, pages 309–316, 2004. DOI: 10.1002/asi.20427 Cited on page(s) 62, 63

S. Corston-Oliver, E. Ringger, M. Gamon, and R. Campbell. Task-focused summarization of email. In *Proc. ACL Workshop Text Summarization Branches Out*, pages 43–50, 2004. Cited on page(s) 71

H.T. Dang. Overview of DUC 2005. In *Proc. 2005 Document Understanding Conference*, pages 1–12, 2005. Cited on page(s) 38

H. Daumé and D. Marcu. Domain adaptation for statistical classifiers. *Journal of Artificial Intelligence Research*, 26(1):101–126, 2006. Cited on page(s) 17, 55

J. Diesner, T. Frantz, and K. Carley. Communication networks from the enron email corpus "It's always about the people. Enron is no different". *Journal of Computational & Mathematical Organization Theory*, 11(3):201–228, 2005. DOI: 10.1007/s10588-005-5377-0 Cited on page(s) 26

B. Dorr, C. Monz, D. Oard, D. Zajic, and R. Schwartz. Extrinsic Evaluation of Automatic Metrics for Summarization. Technical Report LAMP-TR-115,CAR-TR-999,CS-TR-4610,UMIACS-TR-2004-48, University of Maryland, College Park and BBN Technologies*, 2004. DOI: 10.1145/1596517.1596518 Cited on page(s) 33

B. Dorr, C. Monz, S. President, R. Schwartz, and D. Zajic. A methodology for extrinsic evaluation of text summarization: Does ROUGE correlate? In *Proc. ACL Workshop Machine Translation Summarization Evaluation*, pages 1–8, 2005. Cited on page(s) 33, 38

M. Dredze, H. Wallach, D. Puller, and F. Pereira. Generating summary keywords for emails using topics. In *Proc. 13th Int. Conf. on Intelligent User Interfaces*, pages 199–206, 2008. DOI: 10.1145/1378773.1378800 Cited on page(s) 11, 53

M. Elsner and E. Charniak. Disentangling chat. *Computational Linguistics*, 36(3):389–409, 2010. DOI: 10.1162/coli_a_00003 Cited on page(s) 75, 76

B. Endres-Niggemeyer. *Summarizing Information*. Springer, 1998. Cited on page(s) 102

R. Fernandez, M. Frampton, P. Ehlen, M. Purver, and S. Peters. Modelling and detecting decisions in multi-party dialogue. In *Proc. 9th Annual Meeting on Discourse and Dialogue*, pages 156–163, 2008. Cited on page(s) 68

M. Galley. A skip-chain conditional random field for ranking meeting utterances by importance. In *Proc. 2006 Conf. Empirical Methods in Natural Language Processing*, pages 364–372, 2006. DOI: 10.3115/1610075.1610126 Cited on page(s) 36, 37, 91

M. Galley, K. Mckeown, E. Lussier, and H. Jing. Discourse segmentation of multi-party conversation. In *Proc. 41st Annual Meeting Assoc. for Computational Linguistics*, pages 562–569, 2003. DOI: 10.3115/1075096.1075167 Cited on page(s) 50, 51

M. Georgescul, A. Clark, and S. Armstrong. A comparative study of mixture models for automatic topic segmentation of multiparty dialogues. In *Proc. 46th Annual Meeting Assoc. for Computational Linguistics*, pages 925–930, 2008. Cited on page(s) 52

G. Gupta, C. Mazumdar, and M.S. Rao. Digital forensic analysis of e-mails: A trusted e-mail protocol. *International Journal of Digital Evidence*, 2(4):1–11, 2004. Cited on page(s) 8

T. Hain, L. Burget, J. Dines, G. Garau, V. Wan, M. Karafiat, J. Vepa, and M. Lincoln. The AMI system for transcription of speech in meetings. In *Proc. 32nd IEEE Int. Conf. on Acoustics, Speech and Signal Processing*, pages 357–360, 2007. Cited on page(s) 92

S. Havre, E. Hetzler, P. Whitney, and L. Nowell. Themeriver: Visualizing thematic changes in large document collections. *IEEE Trans. on Visualization and Computer Graphics*, 8(1):9–20, 2002. DOI: 10.1109/2945.981848 Cited on page(s) 18

M. Hearst. Texttiling: segmenting text into multi-paragraph subtopic passages. *Computational Linguistics*, 23(1):33–64, 1997. Cited on page(s) 43, 45

Marti A. Hearst. *Search User Interfaces*. Cambridge University Press, 2009. Cited on page(s) 9, 18

D. Heckerman, D. Chickering, C. Meek, R. Rounthwaite, and C. Kadie. Dependency networks for inference, collaborative filtering, and data visualization. *Journal of Machine Learning Research*, 1: 49–75, 2001. DOI: 10.1162/153244301753344614 Cited on page(s) 63

L. Hirschman, M. Light, and E. Breck. Deep read: A reading comprehension system. In *Proc. 27th Annual Meeting Assoc. for Computational Linguistics*, pages 325–332, 1999. DOI: 10.3115/1034678.1034731 Cited on page(s) 38

E. Hovy, C-Y. Lin, and L. Zhou. Evaluating DUC 2005 using basic elements. In *Proc. 2005 Document Understanding Conference*, 2005. Cited on page(s) 33

E. Hovy, C. Y. Lin, L. Zhou, and J. Fukumoto. Automated Summarization Evaluation with Basic Elements. In *Proc. 5th Int. Conf. on Language Resources and Evaluation*, 2006. Cited on page(s) 33

P-Y. Hsueh and J. Moore. What decisions have you made: Automatic decision detection in conversational speech. In *Proc. 8th Annual Meeting North American Assoc. for Computational Linguistics*, pages 25–32, 2007. Cited on page(s) 68

P-Y. Hsueh, J. Moore, and S. Renals. Automatic segmentation of multiparty dialogue. In *Proc. 11th Conf. European Assoc. Computational Linguistics*, pages 273–280, 2006. Cited on page(s) 51

P-Y. Hsueh, J. Kilgour, J. Carletta, J. Moore, and S. Renals. Automatic decision detection in meeting speech. In *Proc. 4th Joint Workshop on Machine Learning and Multimodal Interaction*, pages 168–179, 2007. Cited on page(s) 68

M. Hu, A. Sun, and E. Lim. Comments-oriented blog summarization by sentence extraction. In *Proc. 16th Int. Conf. on Information and Knowledge Management*, pages 901–904, 2007. DOI: 10.1145/1321440.1321571 Cited on page(s) 95, 96

S. Huang and S. Renals. Modeling topic and role information in meetings using the hierarchical dirichlet process. In *Proc. 5th Joint Workshop on Machine Learning and Multimodal Interaction*, pages 214–225, 2008. DOI: 10.1007/978-3-540-85853-9_20 Cited on page(s) 52

A. Janin, D. Baron, J. Edwards, D. Ellis, D. Gelbart, N. Morgan, B. Peskin, T. Pfau, E. Shriberg, A. Stolcke, and C. Wooters. The ICSI meeting corpus. In *Proc. 28th IEEE Int. Conf. on Acoustics, Speech and Signal Processing*, pages 364–367, 2003. DOI: 10.1109/ICASSP.2003.1198793 Cited on page(s) 23, 27

S. Jekat, A. Klein, E. Maier, I. Maleck, M. Mast, and J. Quantz. Dialogue acts in VERBMOBIL. Technical Report VM-Report 65, DFKI, SaarbrÃ¼cken, Germany,, 1995. Cited on page(s) 61

M. Jeong, C-Y. Lin, and G. Lee. Semi-supervised speech act recognition in emails and forums. In *Proc. 2009 Conf. Empirical Methods in Natural Language Processing*, pages 1250–1259, 2009. Cited on page(s) 61, 65

H. Jing, R. Barzilay, K. McKeown, and M. Elhadad. Summarization evaluation methods: Experiments and analysis. In *Proc. 1998 AAAI Symposium on Intelligent Summarization*, pages 60–68, 1998. Cited on page(s) 38

K. Spärck Jones. Automatic summarizing: Factors and directions. In I. Mani and M. Maybury, editors, *Advances in Automatic Text Summarization*, pages 1–12. MITP, 1999. Cited on page(s) 37, 82, 102

K. Spärck Jones and J. Galliers. *Evaluating Natural Language Processing Systems: An Analysis and Review*. Springer, 1995. Cited on page(s) 32

S. Joty, G. Carenini, G. Murray, and R. Ng. Exploiting conversation structure in unsupervised topic segmentation for emails. In *Proc. 2010 Conf. Empirical Methods in Natural Language Processing*, pages 388–398, 2010. Cited on page(s) 55

S. Joty, G. Carenini, and C-Y. Lin. Unsupervised modeling of dialog acts in asynchronous conversations. In *Proc. 22nd Int. Joint Conf. on AI*, 2011. Cited on page(s) 67

D. Jurafsky and J. H. Martin. *Speech and Language Processing*. Prentice Hall, 2008. Cited on page(s) 5, 9, 13, 45, 77, 102

S-M. Kim and E. Hovy. Automatic detection of opinion bearing words and sentences. In *Proc. 2nd Int. Joint Conf. on Natural Language Processing*, pages 61–66, 2005. Cited on page(s) 59

T. Kleinbauer, S. Becker, and T. Becker. Combining multiple information layers for the automatic generation of indicative meeting abstracts. In *Proc. 11th Biennial European Workshop on Natural Language Generation*, pages 151–154, 2007. DOI: 10.3115/1610163.1610190 Cited on page(s) 92, 93

B. Klimt and Y. Yang. Introducing the Enron corpus. In *Proc. 1st Conf. on Email and Anti-Spam*, 2004. Cited on page(s) 26

S. Kübler, R. McDonald, and J. Nivre. *Dependency Parsing*. Morgan & Claypool Publishers, 2009. Cited on page(s) 66

D. Lam, S. Rohall, C. Schmandt, and M. Stern. Exploiting e-mail structure to improve summarization. Technical Report TR2002-02, IBM Research, 2002. Cited on page(s) 13, 86, 91

A. Lampert, R. Dale, and C. Paris. Detecting emails containing requests for action. In *Proc. 11th Annual Meeting North American Assoc. for Computational Linguistics*, pages 984–992, 2010. Cited on page(s) 71

C-Y. Lin. ROUGE: A Package for Automatic Evaluation of Summaries. In *Proc. ACL Workshop Text Summarization Branches Out*, pages 74–81, 2004. Cited on page(s) 33

C. Y. Lin and E. Hovy. Automatic evaluation of summaries using n-gram co-occurrence statistics. In *Proc. 4th Annual Meeting North American Assoc. for Computational Linguistics*, pages 150–156, 2003. DOI: 10.3115/1073445.1073465 Cited on page(s) 32

F. Liu and Y. Liu. Exploring correlation between rouge and human evaluation on meeting summaries. *IEEE Trans. on Audio, Speech & Language Processing*, 18(1):187–196, 2010. DOI: 10.1109/TASL.2009.2025096 Cited on page(s) 19, 33

Y. Liu, F. Liu, B. Li, and S. Xie. Do disfluencies affect meeting summarization: A pilot study on the impact of disfluencies. In *Proc. 4th Joint Workshop on Machine Learning and Multimodal Interaction*, page poster, 2007. Cited on page(s) 91

Y. Liu, S. Xie, and F. Liu. Using N-best recognition output for extractive summarization and keyword extraction in meeting speech. In *Proc. 35th IEEE Int. Conf. on Acoustics, Speech and Signal Processing*, pages 5310–5313, 2010. DOI: 10.1109/ICASSP.2010.5494972 Cited on page(s) 92

I. Mani. *Automatic Summarization.* John Benjamins Co, 2001a. Cited on page(s) 102

I. Mani. Summarization evaluation: An overview. In *Proc. 2nd Workshop on Research in Chinese & Japanese Text Retrieval and Text Summarization*, pages 77–85, 2001b. Cited on page(s) 36, 37, 38, 41

I. Mani and M. Maybury. *Advances in Automatic Text Summarization.* MIT Press, 1999. Cited on page(s) 102

I. Mani, D. House, G. Klein, L. Hirschman, T. Firmin, and B. Sundheim. The TIPSTER SUMMAC text summarization evaluation. In *Proc. 9th Conf. European Assoc. Computational Linguistics*, pages 77–85, 1999. DOI: 10.3115/977035.977047 Cited on page(s) 36, 38

A. McCallum, X. Wang, and A. Corrada-Emmanuel. Topic and role discovery in social networks with experiments on Enron and academic email. *Journal of Artificial Intelligence Research*, 30(1): 249–272, 2007. DOI: 10.1613/jair.2229 Cited on page(s) 78

K. McKeown, R. Barzilay, D. Evans, V. Hatzivassiloglou, J. L. Klavans, A. Nenkova, C. Sable, B. Schiffman, and S. Sigelman. Tracking and summarizing news on a daily basis with Columbia's Newsblaster. In *Proc. 2002 Int. Conf. on Human Language Technology*, pages 280–285, 2002. DOI: 10.3115/1289189.1289212 Cited on page(s) 17

R. Mihalcea and S. Hassan. Using the Essence of Texts to Improve Document Classification. In *Proc. 2005 Int. Conf. on Recent Advances in Natural Language Processing*, pages 1688–1689, 2005. Cited on page(s) 38

G. Mishne and M. de Rijke. Capturing global mood levels using blog posts. In *Proc. 2006 AAAI Symposium on Computational Approaches to Analysing Weblogs*, pages 145–152, 2006. Cited on page(s) 59

G. Mishne and N. Glance. Predicting movie sales from blogger sentiment. In *Proc. 2006 AAAI Symposium on Computational Approaches to Analysing Weblogs*, pages 155–158, 2006. Cited on page(s) 59, 60

A. Morris, G. Kasper, and D. Adams. The effects and limitations of automated text condensing on reading comprehension performance. *Information Systems Research*, 3(1):17–35, 1992. DOI: 10.1287/isre.3.1.17 Cited on page(s) 38

J. Morris and G. Hirst. Lexical cohesion computed by thesaural relations as an indicator of structure of text. *Computational Linguistics*, 17(1):21–48, 1991. Cited on page(s) 50

T. Mullen and R. Malouf. A preliminary investigation into sentiment analysis of informal political discourse. In *Proc. 2006 AAAI Symposium on Computational Approaches to Analysing Weblogs*, pages 159–162, 2006. Cited on page(s) 60

S. Muresan, E. Tzoukermann, and J. Klavans. Combining linguistic and machine learning techniques for email summarization. In *Proc. 5th Int. Conf. on Computational Natural Language Learning*, pages 19:1–19:8, 2001. DOI: 10.3115/1117822.1117837 Cited on page(s) 13, 86

K. Murphy. *Machine Learning: a Probabilistic Perspective*. MIT Press, expected, Spring 2012. Cited on page(s) 9

G. Murray. *Using Speech-Specific Characteristics for Automatic Speech Summarization*. PhD thesis, University of Edinburgh, Edinburgh, Scotland, 2007. Cited on page(s) 93

G. Murray and G. Carenini. Summarizing Spoken and Written Conversations. In *Proc. 2008 Conf. Empirical Methods in Natural Language Processing*, pages 773–782, 2008. DOI: 10.3115/1613715.1613813 Cited on page(s) 17, 98, 100, 101

G. Murray and G. Carenini. Subjectivity detection in spoken and written conversations. *Journal of Natural Language Engineering*, CPO(online), 2010. DOI: 10.1017/S1351324910000264 Cited on page(s) 59

G. Murray and S. Renals. Detecting action items in meetings. In *Proc. 5th Joint Workshop on Machine Learning and Multimodal Interaction*, pages 208–213, 2008. DOI: 10.1007/978-3-540-85853-9_19 Cited on page(s) 70

G. Murray, S. Renals, and J. Carletta. Extractive summarization of meeting recordings. In *Proc. 6th Annual Interspeech Conference*, pages 593–596, 2005a. Cited on page(s) 91, 92

G. Murray, S. Renals, J. Carletta, and J. Moore. Evaluating automatic summaries of meeting recordings. In *Proc. ACL Workshop Machine Translation Summarization Evaluation*, pages 33–40, 2005b. Cited on page(s) 33, 91

G. Murray, S. Renals, J. Moore, and J. Carletta. Incorporating speaker and discourse features into speech summarization. In *Proc. 7th Annual Meeting North American Assoc. for Computational Linguistics*, pages 367–374, 2006. DOI: 10.3115/1220835.1220882 Cited on page(s) 33, 36, 91

G. Murray, T. Kleinbauer, P. Poller, S. Renals, T. Becker, and J. Kilgour. Extrinsic summarization evaluation: A decision audit task. *ACM Trans. on Speech & Language Processing*, 6(2), 2009. DOI: 10.1145/1596517.1596518 Cited on page(s) 7, 38, 81

G. Murray, G. Carenini, and R. Ng. Generating and validating abstracts of meeting conversations: a user study. In *Proc. 6th Biennial Int. Conf. on Natural Language Generation*, pages 105–113, 2010. Cited on page(s) 15, 99, 100, 101

A. Nenkova and A. Bagga. Facilitating email thread access by extractive summary generation. In *Proc. 2003 Int. Conf. on Recent Advances in Natural Language Processing*, pages 287–296, 2003. Cited on page(s) 15, 88, 89, 90

A. Nenkova and B. Passonneau. Evaluating content selection in summarization: The Pyramid method. In *Proc. 5th Annual Meeting North American Assoc. for Computational Linguistics*, pages 145–152, 2004. Cited on page(s) 34, 35

A. Nenkova, R. Passonneau, and K. McKeown. The Pyramid Method: Incorporating Human Content Selection Variation in Summarization Evaluation. *ACM Trans. on Speech & Language Processing*, 4(2):4:1–4:23, 2007. DOI: 10.1145/1233912.1233913 Cited on page(s) 36, 37

P. Newman and J. Blitzer. Summarizing archived discussions: A beginning. In *Proc. 8th Int. Conf. on Intelligent User Interfaces*, pages 273–276, 2003. DOI: 10.1145/604045.604097 Cited on page(s) 87, 89

I. Ounis, C. Macdonald, and I. Soboroff. On the TREC Blog Track. In *Proc. 2nd Int. AAAI Conf. on Weblogs and Social Media*, pages 93–101, 2008. Cited on page(s) 60

B. Pang and L. Lee. Opinion mining and sentiment analysis. *Foundations & Trends in Information Retrieval*, 2(1-2):1–135, 2008. DOI: 10.1561/1500000011 Cited on page(s) 9, 12, 18, 55, 77

K. Papineni, S. Roukos, T. Ward, and W. Zhu. Bleu: a method for automatic evaluation of machine translation. Technical Report RC22176 (W0109-022), IBM Research Division, Thomas J. Watson Research Center, 2001. DOI: 10.3115/1073083.1073135 Cited on page(s) 32

G. Penn and X. Zhu. A critical reassessment of evaluation baselines for speech summarization. In *Proc. 46th Annual Meeting Assoc. for Computational Linguistics*, pages 470–478, 2008. Cited on page(s) 93, 98, 101

G. Penn and X. Zhu. *Speech Summarization*. Morgan & Claypool Publishers, Forthcoming. Cited on page(s) 5, 9, 102

D. Poole and A. Mackworth. *Artificial Intelligence: Foundations of Computational Agents*. Cambridge University Press, 2010. Also full text online at http://artint.info. Cited on page(s) 9, 48, 49, 51, 95

F. Portet, E. Reiter, A. Gatt, J. Hunter, S. Srip ada, Y. Freer, and C. Sykes. Automatic generation of textual summaries from neonatal intens ive care data. *Artificial Intelligence*, 173(7-8):789–816, 2009. DOI: 10.1016/j.artint.2008.12.002 Cited on page(s) 99

M. Purver. Topic segmentation. In G. Tur and R. de Mori, editors, *Spoken Language Understanding: Systems for Extracting Semantic Information from Speech*, pages 291–317. Wiley, 2011. Cited on page(s) 77

M. Purver, P. Ehlen, and J. Niekrasz. Detecting action items in multi-party meetings: Annotation and initial experiments. In *Proc. 3rd Joint Workshop on Machine Learning and Multimodal Interaction*, pages 200–211, 2006a. DOI: 10.1007/11965152_18 Cited on page(s) 70

M. Purver, K. Körding, T. Griffiths, and J. Tenenbaum. Unsupervised topic modelling for multi-party spoken discourse. In *Proc. 44th Annual Meeting Assoc. for Computational Linguistics*, pages 17–24, 2006b. DOI: 10.3115/1220175.1220178 Cited on page(s) 11, 51

M. Purver, J. Dowding, J. Niekrasz, P. Ehlen, and S. Noorbaloochi. Detecting and summarizing action items in multi-party dialogue. In *Proc. 8th Annual Meeting on Discourse and Dialogue*, 2007. Cited on page(s) 70, 71

R. Quirk, S. Greenbaum, G. Leech, and J. Svartvik. *A Comprehensive Grammar of the English Language*. Longman, 1985. Cited on page(s) 25

S. Raaijmakers, K. Truong, and T. Wilson. Multimodal subjectivity analysis of multiparty conversation. In *Proc. 2008 Conf. Empirical Methods in Natural Language Processing*, pages 466–474, 2008. DOI: 10.3115/1613715.1613774 Cited on page(s) 59

D. Radev, H. Jing, M. Styś, and D. Tam. Centroid-based summarization of multiple documents. *Information Processing & Management*, 40(6):919–938, 2004. DOI: 10.1016/j.ipm.2003.10.006 Cited on page(s) 17

D. Ramage, D. Hall, R. Nallapati, and C. Manning. Labeled LDA: A supervised topic model for credit attribution in multi-labeled corpora. In *Proc. 2009 Conf. Empirical Methods in Natural Language Processing*, pages 248–256, 2009. DOI: 10.3115/1699510.1699543 Cited on page(s) 52

D. Ramage, S. Dumais, and D. Liebling. Characterizing microblogs with topic models. In *ICWSM*, pages 130–137, 2010. Cited on page(s) 11, 52, 53

O. Rambow, L. Shrestha, J. Chen, and C. Lauridsen. Summarizing email threads. In *Proc. 5th Annual Meeting North American Assoc. for Computational Linguistics*, pages 105–108, 2004. Cited on page(s) 13, 87, 89, 90

G. Rath, A. Resnick, and R. Savage. The formation of abstracts by the selection of sentences: Part 1: Sentence selection by man and machines. *American Documentation*, 12(2):139–141, 1961. DOI: 10.1002/asi.5090120210 Cited on page(s) 36

E. Reiter and R. Dale. *Building Natural Language Generation Systems.* Cambridge University Press, 2000. Cited on page(s) 85, 99

E. Riloff and W. Phillips. An introduction to the Sundance and AutoSlog systems. Technical Report UUCS-04-015, University of Utah School of Computing, 2004. Cited on page(s) 58, 59

A. Ritter, C. Cherry, and B. Dolan. Unsupervised modeling of twitter conversations. In *Proc. 11th Annual Meeting North American Assoc. for Computational Linguistics*, pages 172–180, 2010. Cited on page(s) 66, 67

O. Sandu, G. Carenini, G. Murray, and R. Ng. Domain adaptation to summarize human conversations. In *Proc. ACL Workshop Domain Adaptation for Natural Language Processing*, pages 16–22, 2010. Cited on page(s) 17, 98

J. Searle. A taxonomy of illocutionary acts. In K. Gunderson, editor, *Minnesota studies in the philosophy of language*, pages 334–369. University of Minnesota Press, 1975. Cited on page(s) 77

T. Segaran, C. Evans, and J. Taylor. *Programming the Semantic Web.* O'Reilly Media, 2009. Cited on page(s) 85

B. Sharifi, M.A. Hutton, and J. Kalita. Summarizing microblogs automatically. In *Proc. 11th Annual Meeting North American Assoc. for Computational Linguistics*, pages 685–688, 2010. Cited on page(s) 8, 15, 95, 96, 97

L. Shrestha and K. McKeown. Detection of question-answer pairs in email conversations. In *Proc. 20th Biennial Int. Conf. on Computational Linguistics*, pages 889–895, 2004. DOI: 10.3115/1220355.1220483 Cited on page(s) 63, 66

E. Shriberg, R. Dhillon, S. Bhagat, J. Ang, , and H. Carvey. The ICSI meeting recorder dialog act (MRDA) corpus. In *Proc. 5th Annual Meeting on Discourse and Dialogue*, pages 97–100, 2004. Cited on page(s) 24, 25

M. Smith, J.J. Cadiz, and B. Burkhalter. Conversation trees and threaded chats. In *Proc. 18th Conf. on Computer Supported Cooperative Work*, pages 97–105, 2000. DOI: 10.1145/358916.358980 Cited on page(s) 75

S. Somasundaran, J. Ruppenhofer, and J. Wiebe. Detecting arguing and sentiment in meetings. In *Proc. 8th Annual Meeting on Discourse and Dialogue*, pages 26–34, 2007. Cited on page(s) 58

M. Steyvers and T. Griffiths. Probabilistic topic models. In T. Landauer, D. Mcnamara, S. Dennis, and W. Kintsch, editors, *Handbook of Latent Semantic Analysis*, pages 427–449. Laurence Erlbaum, 2006. Cited on page(s) 77

A. Stolcke, N. Coccaro, R. Bates, P. Taylor, C. Van Ess-Dykema, K. Ries, E. Shriberg, D. Jurafsky, R. Martin, and M. Meteer. Dialogue act modeling for automatic tagging and recognition of conversational speech. *Computational Linguistics*, 26(3):339–373, 2000. DOI: 10.1162/089120100561737 Cited on page(s) 23

C. Sutton and A. McCallum. Collective segmentation and labeling of distant entities in information extraction. Technical Report 04-49, University of Massachusetts, 2004. Cited on page(s) 91

M. Taboada, J. Brooke, M. Tofiloski, K. Voll, and M. Stede. Lexicon-based methods for sentiment analysis. *Computational Linguistics*, 37(2):267–307, 2010. DOI: 10.1162/COLI_a_00049 Cited on page(s) 57

L. Tang and H. Liu. *Community Detection and Mining in Social Media.* Morgan & Claypool Publishers, 2010. DOI: 10.2200/S00298ED1V01Y201009DMK003 Cited on page(s) 78

S. Teufel and H. van Halteren. Evaluating information content by factoid analysis: Human annotation and stability. In *Proc. 2004 Conf. Empirical Methods in Natural Language Processing*, pages 419–426, 2004. Cited on page(s) 37

S. Tucker and S. Whittaker. Time is of the essence: an evaluation of temporal compression algorithms. In *Proc. 24th Int. SIGCHI Conf. on Human Factors in Computing Systems*, pages 329–338, 2006. DOI: 10.1145/1124772.1124822 Cited on page(s) 94

J. Ulrich, G. Murray, and G. Carenini. A publicly available annotated corpus for supervised email summarization. In *Proc. AAAI Workshop on Enhanced Messaging*, pages 77–82, 2008. Cited on page(s) 26

M. Utiyama and H. Isahara. A statistical model for domain-independent text segmentation. In *Proc. 39th Annual Meeting Assoc. for Computational Linguistics*, pages 499–506, 2001. DOI: 10.3115/1073012.1073076 Cited on page(s) 45

A. Waibel, M. Bett, M. Finke, and R. Stiefelhagen. Meeting browser: tracking and summarization meetings. In *Proc. 1998 DARPA Workshop on Broadcast News*, pages 281–286, 1998. Cited on page(s) 91

S. Wan and K. McKeown. Generating overview summaries of ongoing email thread discussions. In *Proc. 20th Biennial Int. Conf. on Computational Linguistics*, pages 549–555, 2004. DOI: 10.3115/1220355.1220434 Cited on page(s) 59, 87, 88, 89

Y-C. Wang, M. Joshi, W. Cohen, and C. Rose. Recovering Implicit Thread Structure in Newsgroup Style Conversations. In *Proc. 2nd Int. AAAI Conf. on Weblogs and Social Media*, 2008. Cited on page(s) 78

T. Wilson. Annotating subjective content in meetings. In *Proc. 6th Int. Conf. on Language Resources and Evaluation*, pages 2738–2745, 2008. Cited on page(s) 25, 27

T. Wilson, P. Hoffman, S. Somasundaran, J. Kessler, J. Wiebe, Y. Choi, C. Cardie, E. Riloff, and S. Patwardhan. OpinionFinder: A System for Subjectivity Analysis. In *Proc. 2005 Conf. Empirical Methods in Natural Language Processing*, pages 34–35, 2005. Cited on page(s) 59

T. Wilson, J. Wiebe, and R. Hwa. Recognizing strong and weak opinion clauses. *Computational Intelligence*, 22(2):73–99, 2006. DOI: 10.1111/j.1467-8640.2006.00275.x Cited on page(s) 12

J. Yeh and A. Harnly. Email thread reassembly using similarity matching. In *Proc. 3rd Conf. on Email and Anti-Spam*, 2006. Cited on page(s) 26

K. Zechner. Automatic summarization of open-domain multiparty dialogues in diverse genres. *Computational Linguistics*, 28(4):447–485, 2002. DOI: 10.1162/089120102762671945 Cited on page(s) 97

K. Zechner and A. Waibel. Minimizing word error rate in textual summaries of spoken language. In *Proc. 1st Annual Meeting North American Assoc. for Computational Linguistics*, pages 186–193, 2000. Cited on page(s) 37

L. Zhou and E. Hovy. Digesting virtual "geek" culture: The summarization of technical internet relay chats. In *Proc. 43rd Annual Meeting Assoc. for Computational Linguistics*, pages 298–305, 2005. DOI: 10.3115/1219840.1219877 Cited on page(s) 95, 96

L. Zhou and E. Hovy. On the summarization of dynamically introduced information: Online discussions and blogs. In *Proc. 2006 AAAI Symposium on Computational Approaches to Analysing Weblogs*, pages 237–242, 2006. Cited on page(s) 60

L. Zhou, M. Ticrea, and E. Hovy. Multi-document biography summarization. In *Proc. 2004 Conf. Empirical Methods in Natural Language Processing*, 2004. Cited on page(s) 17

X. Zhu. Semi-supervised learning literature survey. Technical Report 1530, University of Wisconsin at Madison, Department of Computer Sciences, 2006. Cited on page(s) 92

X. Zhu and A. Goldberg. *Introduction to Semi-Supervised Learning*. Morgan & Claypool Publishers, 2009. DOI: 10.2200/S00196ED1V01Y200906AIM006 Cited on page(s) 9

Authors' Biographies

GIUSEPPE CARENINI

Dr. Giuseppe Carenini is an associate professor in computer science at UBC, with broad interdisciplinary interests. His work on combining natural language processing and information visualization to support decision making has been published in over 70 peer-reviewed papers. Dr. Carenini was the area chair for "Sentiment Analysis, Opinion Mining, and Text Classification" of ACL 2009 and he is currently co-editing an ACM TIST Special Issue on Intelligent Visual Interfaces for Text Analysis. In his work, Dr. Carenini has also extensively collaborated with industrial partners, including Microsoft and IBM.

GABRIEL MURRAY

Dr. Gabriel Murray is a researcher in computer science at UBC, in the Laboratory for Computational Intelligence. His background is in natural language processing as well as theoretical linguistics. He has an established research record in the area of automatic summarization, with particular attention to summarization of noisy genres such as speech and web data, and comparison of abstractive and extractive techniques. He did his graduate studies at the University of Edinburgh under Dr. Steve Renals, and was a member of the EU-funded AMI project on studying multimodal interaction. He is currently a researcher with the NSERC Business Intelligence Network on intelligent data management and decision making.

RAYMOND NG

Dr. Raymond Ng is a professor in computer science at UBC. He is internationally renowned for his data mining studies. He has published over 100 journal and conference papers covering a broad range of topics in informatics, data mining and databases. He has won Best Paper awards from the ACM SIGKDD conference on data mining and the ACM SIGMOD conference on database management. For the past few years, Dr. Ng has been one of the editors of two top database journals worldwide—the VLDB Journal and the IEEE Transactions on Knowledge and Data Engineering. He was the general chair of ACM SIGMOD 2008 and the program chair of IEEE ICDE 2009.